LONDON MATHEMATICAL SOCIETY LECTURE

Managing Editor: Professor M. Reid, Mathematics Institute,
University of Warwick, Coventry CV4 7AL, United Kingdom

The titles below are available from booksellers, or from Cambridg
http://www.cambridge.org/mathematics

333 Synthetic differential geometry (2nd Edition), A. KOCK
334 The Navier–Stokes equations, N. RILEY & P. DRAZIN
335 Lectures on the combinatorics of free probability, A. NICA & R. SPEICHER
336 Integral closure of ideals, rings, and modules, I. SWANSON & C. HUNEKE
337 Methods in Banach space theory, J.M.F. CASTILLO & W.B. JOHNSON (eds)
338 Surveys in geometry and number theory, N. YOUNG (ed)
339 Groups St Andrews 2005 I, C.M. CAMPBELL, M.R. QUICK, E.F. ROBERTSON & G.C. SMITH (eds)
340 Groups St Andrews 2005 II, C.M. CAMPBELL, M.R. QUICK, E.F. ROBERTSON & G.C. SMITH (eds)
341 Ranks of elliptic curves and random matrix theory, J.B. CONREY, D.W. FARMER, F. MEZZADRI & N.C. SNAITH (eds)
342 Elliptic cohomology, H.R. MILLER & D.C. RAVENEL (eds)
343 Algebraic cycles and motives I, J. NAGEL & C. PETERS (eds)
344 Algebraic cycles and motives II, J. NAGEL & C. PETERS (eds)
345 Algebraic and analytic geometry, A. NEEMAN
346 Surveys in combinatorics 2007, A. HILTON & J. TALBOT (eds)
347 Surveys in contemporary mathematics, N. YOUNG & Y. CHOI (eds)
348 Transcendental dynamics and complex analysis, P.J. RIPPON & G.M. STALLARD (eds)
349 Model theory with applications to algebra and analysis I, Z. CHATZIDAKIS, D. MACPHERSON, A. PILLAY & A. WILKIE (eds)
350 Model theory with applications to algebra and analysis II, Z. CHATZIDAKIS, D. MACPHERSON, A. PILLAY & A. WILKIE (eds)
351 Finite von Neumann algebras and masas, A.M. SINCLAIR & R.R. SMITH
352 Number theory and polynomials, J. MCKEE & C. SMYTH (eds)
353 Trends in stochastic analysis, J. BLATH, P. MÖRTERS & M. SCHEUTZOW (eds)
354 Groups and analysis, K. TENT (ed)
355 Non-equilibrium statistical mechanics and turbulence, J. CARDY, G. FALKOVICH & K. GAWEDZKI
356 Elliptic curves and big Galois representations, D. DELBOURGO
357 Algebraic theory of differential equations, M.A.H. MACCALLUM & A.V. MIKHAILOV (eds)
358 Geometric and cohomological methods in group theory, M.R. BRIDSON, P.H. KROPHOLLER & I.J. LEARY (eds)
359 Moduli spaces and vector bundles, L. BRAMBILA-PAZ, S.B. BRADLOW, O. GARCÍA-PRADA & S. RAMANAN (eds)
360 Zariski geometries, B. ZILBER
361 Words: Notes on verbal width in groups, D. SEGAL
362 Differential tensor algebras and their module categories, R. BAUTISTA, L. SALMERÓN & R. ZUAZUA
363 Foundations of computational mathematics, Hong Kong 2008, F. CUCKER, A. PINKUS & M.J. TODD (eds)
364 Partial differential equations and fluid mechanics, J.C. ROBINSON & J.L. RODRIGO (eds)
365 Surveys in combinatorics 2009, S. HUCZYNSKA, J.D. MITCHELL & C.M. RONEY-DOUGAL (eds)
366 Highly oscillatory problems, B. ENGQUIST, A. FOKAS, E. HAIRER & A. ISERLES (eds)
367 Random matrices: High dimensional phenomena, G. BLOWER
368 Geometry of Riemann surfaces, F.P. GARDINER, G. GONZÁLEZ-DIEZ & C. KOUROUNIOTIS (eds)
369 Epidemics and rumours in complex networks, M. DRAIEF & L. MASSOULIÉ
370 Theory of p-adic distributions, S. ALBEVERIO, A.YU. KHRENNIKOV & V.M. SHELKOVICH
371 Conformal fractals, F. PRZYTYCKI & M. URBAŃSKI
372 Moonshine: The first quarter century and beyond, J. LEPOWSKY, J. MCKAY & M.P. TUITE (eds)
373 Smoothness, regularity and complete intersection, J. MAJADAS & A. G. RODICIO
374 Geometric analysis of hyperbolic differential equations: An introduction, S. ALINHAC
375 Triangulated categories, T. HOLM, P. JØRGENSEN & R. ROUQUIER (eds)
376 Permutation patterns, S. LINTON, N. RUŠKUC & V. VATTER (eds)
377 An introduction to Galois cohomology and its applications, G. BERHUY
378 Probability and mathematical genetics, N. H. BINGHAM & C. M. GOLDIE (eds)
379 Finite and algorithmic model theory, J. ESPARZA, C. MICHAUX & C. STEINHORN (eds)
380 Real and complex singularities, M. MANOEL, M.C. ROMERO FUSTER & C.T.C WALL (eds)
381 Symmetries and integrability of difference equations, D. LEVI, P. OLVER, Z. THOMOVA & P. WINTERNITZ (eds)
382 Forcing with random variables and proof complexity, J. KRAJÍČEK
383 Motivic integration and its interactions with model theory and non-Archimedean geometry I, R. CLUCKERS, J. NICAISE & J. SEBAG (eds)
384 Motivic integration and its interactions with model theory and non-Archimedean geometry II, R. CLUCKERS, J. NICAISE & J. SEBAG (eds)
385 Entropy of hidden Markov processes and connections to dynamical systems, B. MARCUS, K. PETERSEN & T. WEISSMAN (eds)
386 Independence-friendly logic, A.L. MANN, G. SANDU & M. SEVENSTER

387 Groups St Andrews 2009 in Bath I, C.M. CAMPBELL *et al* (eds)
388 Groups St Andrews 2009 in Bath II, C.M. CAMPBELL *et al* (eds)
389 Random fields on the sphere, D. MARINUCCI & G. PECCATI
390 Localization in periodic potentials, D.E. PELINOVSKY
391 Fusion systems in algebra and topology, M. ASCHBACHER, R. KESSAR & B. OLIVER
392 Surveys in combinatorics 2011, R. CHAPMAN (ed)
393 Non-abelian fundamental groups and Iwasawa theory, J. COATES *et al* (eds)
394 Variational problems in differential geometry, R. BIELAWSKI, K. HOUSTON & M. SPEIGHT (eds)
395 How groups grow, A. MANN
396 Arithmetic differential operators over the p-adic integers, C.C. RALPH & S.R. SIMANCA
397 Hyperbolic geometry and applications in quantum chaos and cosmology, J. BOLTE & F. STEINER (eds)
398 Mathematical models in contact mechanics, M. SOFONEA & A. MATEI
399 Circuit double cover of graphs, C.-Q. ZHANG
400 Dense sphere packings: a blueprint for formal proofs, T. HALES
401 A double Hall algebra approach to affine quantum Schur–Weyl theory, B. DENG, J. DU & Q. FU
402 Mathematical aspects of fluid mechanics, J.C. ROBINSON, J.L. RODRIGO & W. SADOWSKI (eds)
403 Foundations of computational mathematics, Budapest 2011, F. CUCKER, T. KRICK, A. PINKUS & A. SZANTO (eds)
404 Operator methods for boundary value problems, S. HASSI, H.S.V. DE SNOO & F.H. SZAFRANIEC (eds)
405 Torsors, étale homotopy and applications to rational points, A.N. SKOROBOGATOV (ed)
406 Appalachian set theory, J. CUMMINGS & E. SCHIMMERLING (eds)
407 The maximal subgroups of the low-dimensional finite classical groups, J.N. BRAY, D.F. HOLT & C.M. RONEY-DOUGAL
408 Complexity science: the Warwick master's course, R. BALL, V. KOLOKOLTSOV & R.S. MACKAY (eds)
409 Surveys in combinatorics 2013, S.R. BLACKBURN, S. GERKE & M. WILDON (eds)
410 Representation theory and harmonic analysis of wreath products of finite groups, T. CECCHERINI-SILBERSTEIN, F. SCARABOTTI & F. TOLLI
411 Moduli spaces, L. BRAMBILA-PAZ, O. GARCÍA-PRADA, P. NEWSTEAD & R.P. THOMAS (eds)
412 Automorphisms and equivalence relations in topological dynamics, D.B. ELLIS & R. ELLIS
413 Optimal transportation, Y. OLLIVIER, H. PAJOT & C. VILLANI (eds)
414 Automorphic forms and Galois representations I, F. DIAMOND, P.L. KASSAEI & M. KIM (eds)
415 Automorphic forms and Galois representations II, F. DIAMOND, P.L. KASSAEI & M. KIM (eds)
416 Reversibility in dynamics and group theory, A.G. O'FARRELL & I. SHORT
417 Recent advances in algebraic geometry, C.D. HACON, M. MUSTAŢĂ & M. POPA (eds)
418 The Bloch–Kato conjecture for the Riemann zeta function, J. COATES, A. RAGHURAM, A. SAIKIA & R. SUJATHA (eds)
419 The Cauchy problem for non-Lipschitz semi-linear parabolic partial differential equations, J.C. MEYER & D.J. NEEDHAM
420 Arithmetic and geometry, L. DIEULEFAIT *et al* (eds)
421 O-minimality and Diophantine geometry, G.O. JONES & A.J. WILKIE (eds)
422 Groups St Andrews 2013, C.M. CAMPBELL *et al* (eds)
423 Inequalities for graph eigenvalues, Z. STANIĆ
424 Surveys in combinatorics 2015, A. CZUMAJ *et al* (eds)
425 Geometry, topology and dynamics in negative curvature, C.S. ARAVINDA, F.T. FARRELL & J.-F. LAFONT (eds)
426 Lectures on the theory of water waves, T. BRIDGES, M. GROVES & D. NICHOLLS (eds)
427 Recent advances in Hodge theory, M. KERR & G. PEARLSTEIN (eds)
428 Geometry in a Fréchet context, C. T. J. DODSON, G. GALANIS & E. VASSILIOU
429 Sheaves and functions modulo p, L. TAELMAN
430 Recent progress in the theory of the Euler and Navier-Stokes equations, J.C. ROBINSON, J.L. RODRIGO, W. SADOWSKI & A. VIDAL-LÓPEZ (eds)
431 Harmonic and subharmonic function theory on the real hyperbolic ball, M. STOLL
432 Topics in graph automorphisms and reconstruction (2nd Edition), J. LAURI & R. SCAPELLATO
433 Regular and irregular holonomic D-modules, M. KASHIWARA & P. SCHAPIRA
434 Analytic semigroups and semilinear initial boundary value problems (2nd Edition), K. TAIRA
435 Graded rings and graded Grothendieck groups, R. HAZRAT
436 Groups, graphs and random walks, T. CECCHERINI-SILBERSTEIN, M. SALVATORI & E. SAVA-HUSS (eds)
437 Dynamics and analytic number theory, D. BADZIAHIN, A. GORODNIK & N. PEYERIMHOFF (eds)
438 Random walks and heat kernels on graphs, M.T. BARLOW
439 Evolution equations, K. AMMARI & S. GERBI (eds)
440 Surveys in combinatorics 2017, A. CLAESSON *et al* (eds)
441 Polynomials and the mod 2 Steenrod algebra I, G. WALKER & R.M.W. WOOD
442 Polynomials and the mod 2 Steenrod algebra II, G. WALKER & R.M.W. WOOD
443 Asymptotic analysis in general relativity, T. DAUDÉ, D. HÄFNER & J.-P. NICOLAS (eds)
444 Geometric and cohomological group theory, P.H. KROPHOLLER, I.J. LEARY, C. MARTÍNEZ-PÉREZ & B.E.A. NUCINKIS (eds)
445 Introduction to hidden semi-Markov models, J. VAN DER HOEK & R.J. ELLIOTT
446 Advances in two-dimensional homotopy and combinatorial group theory, W. METZLER & S. ROSEBROCK (eds)
447 New directions in locally compact groups, P.-E. CAPRACE & N. MONOD (eds)

London Mathematical Society Lecture Note Series: 445

Introduction to Hidden Semi-Markov Models

JOHN VAN DER HOEK
University of South Australia

ROBERT J. ELLIOTT
University of South Australia and University of Calgary

CAMBRIDGE
UNIVERSITY PRESS

University Printing House, Cambridge CB2 8BS, United Kingdom

One Liberty Plaza, 20th Floor, New York, NY 10006, USA

477 Williamstown Road, Port Melbourne, VIC 3207, Australia

314–321, 3rd Floor, Plot 3, Splendor Forum, Jasola District Centre,
New Delhi – 110025, India

79 Anson Road, #06–04/06, Singapore 079906

Cambridge University Press is part of the University of Cambridge.

It furthers the University's mission by disseminating knowledge in the pursuit of
education, learning, and research at the highest international levels of excellence.

www.cambridge.org
Information on this title: www.cambridge.org/9781108441988
DOI: 10.1017/9781108377423

First published 2018

Printed in the United Kingdom by Clays, St Ives plc

A catalogue record for this publication is available from the British Library.

ISBN 978-1-108-44198-8 Paperback

Contents

Preface

The purpose of this volume is to present the theory of Markov and semi-Markov processes in a discrete-time, finite-state framework. Given this background, hidden versions of these processes are introduced and related estimation and filtering results developed. The approach is similar to the earlier book, Elliott et al. (1995). That is, a central tool is the Radon–Nikodym theorem and related changes of probability measure. In the discrete-time, finite-state framework that we employ these have simple interpretations following from Bayes' theorem.

Markov chains and hidden Markov chains have found many applications in fields from finance, where the chains model different economic regimes, to genomics, where gene and protein structure is modelled as a hidden Markov model. Semi-Markov chains and hidden semi-Markov chains will have similar, possibly more realistic, applications. The genomics applications are modelled by discrete observations of these hidden chains.

Recent books in the area include in particular Koski (2001) and Barbu and Limnios (2008). Koski includes many examples, not much theory and little on semi-Markov Models. Barbu and Limnios say that the estimation of discrete-time semi-Markov systems is almost absent from the literature. They present an alternative specification from the one adopted in this book and so we give alternative methods in a rigorous framework. They provide limited applications in genomics.

This book carefully constructs relevant processes and proves required results. The filters and related parameter estimation methods we obtain for semi-Markov chains include new results. The occupation times in any state of a Markov chain are geometrically distributed; semi-Markov chains can have occupation times which are quite general and not necessarily geometrically distributed.

Works on semi-Markov processes include Barbu and Limnios (2008), Çinlar (1975), Harlamov (2008), Howard (1971), Janssen and Manca (2010), and Koski (2001) from Chapter 11 onwards. Çinlar (1975) considers a countable state space.

Hidden Markov models have found extensive applications in speech processing and genomics. References for these applications include Ferguson (1980), who considers more general occupation times. This problem was also investigated by Levinson (1986a,b), Ramesh and Wilpon (1992), and in the papers Guédon (1992) and Guédon and Cocozza-Thivent (1990). Genomic applications are treated in the thesis of Burge (1997) and the book Burge and Karlin (1997). Applications in financial modelling can be found, for example in the works Bulla (2006), Bulla and Bulla (2006), Bulla et al. (2010), but these use continuous observations, which we do not focus on here.

The book commences with a construction of finite-state Markov chains in discrete time. Filtering results for hidden Markov chains are then established, including a proof of the Viterbi algorithm. In the second part of the book semi-Markov chains are defined followed by hidden semi-Markov chains and related filtering and estimation results, some of which are new.

Developed in the simple discrete-time, finite-state setting the book will provide graduate, and advanced undergraduate readers, with the modern tools and terminology to develop and apply these models. Appendix E outlines applications in genomics.

The contents should be accessible to a reader who has some familiarity with elementary, discrete probability theory. Consequently it is suitable for senior undergraduate and Masters level courses.

Acknowledgements Robert Elliott wishes to thank NSERC and the ARC for continuing support.

1
Observed Markov Chains

1.1 Introduction

This book studies finite state processes in discrete time. The simplest such process is just a sequence of independent random variables which at each time takes any one of the possible values in its state space with equal probability. The canonical probability space for such a process is the space of sequences of outcomes. The following chapter starts by describing this probability space and constructing on it an appropriate measure. A modified construction then gives a probability space on which not all outcomes have equal probability. A Radon–Nikodym derivative is then defined so the sequence of outcomes is no longer independent, rather the probability of the following state depends on the present state. That is, we construct a Markov chain. This construction of a Markov chain from first principles is not given in other treatments of the subject. The semi-martingale representation of the chain is also given.

The following first few chapters contruct Markov chains and hidden Markov chains from first principles. Estimation algorithms are derived. Semi-Markov chains and hidden semi-Markov chains are introduced and discussed from Chapter 9 onwards.

1.2 Observed Markov chain models

Suppose $\{X_k\,; k = 1, 2, \dots\}$ is a sequence of random quantities taking values in some set $\mathcal{S}$.

We say $\{X_k\,; k = 1, 2, \dots, L\}$ is a *Markov chain* if the following properties hold:

$$P(X_k = x_k | X_1 = x_1, X_2 = x_2, \ldots, X_{k-1} = x_{k-1})$$
$$= P(X_k = x_k | X_{k-1} = x_{k-1})$$

for each $k \geq 1$ and for all $x_1, x_2, \ldots, x_k$.

This is also called an M1 model. The iid model is called an M0 model. In an Mq model, a Markov chain of order q:

$$P(X_k = x_k | X_1 = x_1, X_2 = x_2, \ldots, X_{k-1} = x_{k-1})$$
$$= P(X_k = x_k | X_{k-q} = x_{k-q}, \ldots, X_{k-1} = x_{k-1})$$

for each $n \geq 1$ and for all $x_1, x_2, \ldots, x_k$.

It can be shown that a Markov chain of higher order can be reduced to a 1-step Markov chain, so they are of limited independent interest, at least theoretically.

In fact for an $M2$ chain $\{X_n\}$, where we have

$$P(X_k = x_k | X_1 = x_1, \ldots, X_{k-1} = x_{k-1})$$
$$= P(X_k = x_k | X_{k-2} = x_{k-2}, X_{k-1} = x_{k-1}),$$

we obtain an M1 chain if we set

$$Y_k = \begin{pmatrix} X_k \\ X_{k-1} \end{pmatrix}$$

for $k = 2, 3, \ldots, L$.

In our models, the state space S, (the set of values that each term of the chain can take), is finite corresponding to the number of elements in an alphabet. If S has N elements, it is convenient to let S consist of the N standard unit vectors e_i, $i = 1, 2, \ldots, N$, in $\mathbb{R}^N$. Here $e_i = (0, \ldots, 0, 1, 0, \ldots, 0) \in \mathbb{R}^N$. Then the elements of $S = \{e_1, \ldots, e_N\}$ are in one-to-one correspondence with an alphabet $\mathcal{Q}$ having N elements.

From now on we assume $\mathcal{S} = \{e_1, e_2, \ldots, e_N\}$. Associated with a time-homogeneous Markov chain we have well-defined transition probabilities:

$$p_{ij} \equiv p_{e_i, e_j} = P(X_k = e_j | X_{k-1} = e_i).$$

These have the same values for each $k = 2, 3, \ldots, L$. (That is what homogeneous means here.) This is the convention used by many probabilists. However, shall use the convention used by those who work with HMMs and write

$$a_{ji} = P(X_k = e_j | X_{k-1} = e_i).$$

For the matrix (p_{ij}) the row sums are all 1, while for the matrix (a_{ji}) the column sums are all 1.

Along with the transition probabilities, we also need initial probabilities

$$\pi_j \equiv \pi_{e_j} = P(X_1 = e_j)$$

for each $e_j \in S$.

We can then write down the probability of any sample path

$$\mathbf{x} = (x_1, x_2, \ldots, x_L)$$

of

$$\mathbf{X} = (X_1, X_2, \ldots, X_L)$$

as

$$P(\mathbf{x}) \equiv P(\mathbf{X} = \mathbf{x}) = \pi_{x_1} \times \prod_{k=1}^{L} p_{x_{k-1}, x_k} \, .$$

To calibrate this kind of model, we need some 'training data'. This could mean that we have M sequences of length L as sample paths from our model. We could then make estimates

$$\widehat{\pi}_j = \frac{\text{number of times } e_j \text{ occurs as } X_1}{M}$$

and

$$\widehat{a}_{ji} = \widehat{p}_{ij} = \frac{\text{number of times } e_j \text{ follows } e_i}{\text{number of times anything follow } e_i} \, .$$

We shall show that expressions like these are maximum likelihood estimators of these quantities.

In many other applications, there is often only one observed sample path of a Markov chain. In other words, when k of X_k represents time, we only have one observation history. It is not the case that different spectators in this world see different histories (even though they may report events as if that were the case). One observed sequence is the case with financial and economic data, or other tracking signals. In genomics we can often have more than one sample path from the same model. These could be obtained from a DNA molecule by selecting out several subsequences of length L.

In this section, we shall discuss the construction and estimation of observed Markov chain models.

We shall consider a Markov chain X taking values in a finite set $\mathcal{S}$. We have not specified $\mathcal{S}$, except that it has N elements, say. It does not really matter what the objects in $\mathcal{S}$ are as long as we know how to put them in one-to-one correspondence with some alphabet.

As above it is most convenient to identity the elements of $\mathcal{S}$ with the standard unit vectors $e_1, e_2, \ldots, e_N$ in $\mathbb{R}^N$. This means that

$$e_i = (0, 0, \ldots, 1, 0, \ldots, 0)^\top$$

where the 1 is in the ith place and $\top$ denotes the transpose.

We shall first construct a Markov chain $\{X_k; k = 0, 1, 2, \ldots\}$ taking values in a finite state space $\mathcal{S} = \{e_1, e_2, \ldots, e_N\}$. This Markov chain will be defined on a canonical probability space $(\Omega, \mathcal{F}, P)$, which we shall now describe.

Note that with this notation we have the representations:

$$1 = \sum_{i=1}^{N} \langle X_k, e_i \rangle$$

and for any real-valued functions $g(X_k)$

$$g(X_k) = \langle g, X_k \rangle$$

where $g = (g_1, g_2, \ldots, g_N)$ and $g_i = g(e_i)$.

For $u, v \in \mathbb{R}^N$ write $\langle u, v \rangle = u_1 v_1 + \cdots + u_N v_N$, the usual inner product of $\mathbb{R}^N$. We have just used this notation and will continue to use this notation.

1.3 Notation

We introduce some notation to be used in this section.

The sample space Ω will consist of all sequences of

$$\omega = (\omega_0, \omega_1, \omega_2, \ldots)$$

where $\omega_i \in \mathcal{S}$ for each $i \geq 0$.

A σ-algebra on Ω is a family of subsets $\mathcal{F}$ of Ω which satisfies:

(1) $\Omega \in \mathcal{F}$;
(2) if $A \in \mathcal{F}$ then the complement $A^c \in \mathcal{F}$;
(3) if $A_1, A_2, \ldots$ are all in $\mathcal{F}$ then $\bigcup_{i=1}^{\infty} A_i \in \mathcal{F}$.

Consider the family $\mathcal{F}^A$ of subsets of Ω of the form:

$$\{\omega \in \Omega \,|\, \omega_{i_k} = e_{i_k}, k = 1, 2, \ldots, l\}, \tag{1.1}$$

where $i_1 < i_2 < \cdots < i_l$ and $e_{i_1}, e_{i_2}, \ldots, e_{i_k}$ are elements of $\mathcal{S}$.

The σ-algebra $\mathcal{F}$ we shall consider on Ω will be the smallest σ-algebra generated by all the sets in $\mathcal{F}^A$.

Elements in $\mathcal{F}$ will be called *events*.

Once we have assigned a probability $P(B)$ to each event $B \in \mathcal{F}^A$, then it can be extended to an event of $F \in \mathcal{F}$, by first expressing F as a disjoint union of such sets:

$$F = \bigcup_{i=1}^{\infty} A_i\,, \quad A_i \in \mathcal{F}^A, \text{ with } A_i \cap A_j = \emptyset \text{ for } i \neq j$$

and then defining $P(F)$ by

$$P(F) = \sum_{i=1}^{\infty} P(A_i).$$

There are many ways that a probability can be assigned to an event.

The probability function has the defining properties:

(i) $P(\Omega) = 1$,
(ii) $P(A^c) = 1 - P(A)$, where $A^c = \Omega \setminus A$ for any $A \in \mathcal{F}$,
(iii) if $\{A_n\} \subset \mathcal{F}$ are disjoint, then

$$P\left(\bigcup_{n=1}^{\infty} A_n\right) = \sum_{n=1}^{\infty} P(A_n)\,.$$

The canonical process The canonical process $\{X_k\}$ is defined on Ω by

$$X_k(\omega) = \omega_k \quad \text{for each } \omega \in \Omega$$

for $k = 0, 1, 2, \ldots$ The statistical properties of $\{X_k\}$ will depend on the probability P defined on $\mathcal{F}$.

We let $\mathcal{F}_n \subset \mathcal{F}$ be the collection all subsets of Ω generated by the events A with $i_k \leq n$ in (1.1) for $n = 0, 1, 2, \ldots$. Then we have

$$\Omega = \bigcup_{A \in \mathcal{F}_n} A\,.$$

Note that $\mathcal{F}_n$ is the σ-algebra generated by $X_0, X_1, \ldots, X_n$. This means that knowing the elements of $\mathcal{F}_n$ is equivalent to knowing $X_0, X_1, \ldots, X_n$. The increasing family of σ-algebras $\{\mathcal{F}_n\}$ is called a *filtration* on Ω.

We shall call $\{X_n\}$ a *Markov chain* if it has the following property:

$$P(X_{n+1} = e_j \,|\, \mathcal{F}_n) = P(X_{n+1} = e_j \,|\, X_n).$$

Here the left-hand side is a conditional probability depending on the entire past history of the process $\{X_k \mid k = 0, 1, \ldots, n\}$ while on the right-hand side the conditional probability depends only on the knowledge of X_n.

This implies that we can define transition probabilities

$$a_{ji} = P(X_{n+1} = e_j \mid X_n = e_i)$$

where

$$\sum_{j=1}^{n} a_{ji} = 1\,.$$

This is the case for a (time-)homogeneous Markov chain, as the matrix of probabilities (a_{ji}) does not depend on n. However, the transition probabilities could depend on n so

$$a_{ji}(n) = P(X_{n+1} = e_j | X_n = e_i).$$

Many of the results below extend to the situation. We would then write $a_{ji}(n)$ in place of a_{ji} for all i, j.

As noted earlier some probabilists write

$$p_{ij} = P(X_{n+1} = e_j \mid X_n = e_i).$$

However, we shall not follow this practice as there are some distinct advantages using the above notation which is that used in Elliott et al. (1995) and in other papers.

It is also possible to define Markov chains of higher-order $M \geq 2$. The Markov chain we have just described is the usual one and has order 1. For an order-2 chain we would instead have the condition

$$P(X_{n+1} = e_j \mid \mathcal{F}_n) = P(X_{n+1} = e_j \mid X_n,\, X_{n-1}).$$

As these higher-order Markov chains are used in genomic modelling, we shall describe their representation as an order-1 Markov chain with an extended state space.

1.4 Construction of Markov chains

The reference model We say that we have the reference model when the probability is specified by

$$\overline{P}(B) = \frac{1}{N^l}$$

for events $B \in \mathcal{F}^A$, of the form (1.1).

We shall write $\overline{P}$ and $\overline{\mathbf{E}}$ to indicate probabilities and expectations using this probability.

Properties of the reference model

Property 1: We have

$$\begin{aligned}\overline{P}(X_k = e_j) &\equiv \overline{P}(\{\omega \in \Omega \,|\, X_k(\omega) = e_j\}) = \overline{P}(\{\omega \in \Omega |\, \omega_k = e_j\})\\ &= \frac{1}{N}\end{aligned}$$

for each k and e_j.

This means that each X_k has the same distribution, and this is the uniform distribution, assigning equal probabilities to the occurrence of each state in $\mathcal{S}$.

Property 2: The terms of $\{X_k\}$ are independent.

To show this let $k < l$. Then

$$\begin{aligned}\overline{P}(X_k = e_j, X_l = e_i) &\equiv \overline{P}(\{\omega \in \Omega |\, X_k(\omega) = e_j, X_l(\omega) = e_i\})\\ &= \overline{P}(\{\omega \in \Omega |\, \omega_k = e_j, \omega_l = e_i\})\\ &= \frac{1}{N^2}\\ &= \overline{P}(X_k = e_j)\, \overline{P}(X_l = e_i)\end{aligned}$$

This means that X_k and X_l are independent for any k, l and so the sequence $\{X_n\}$ is a uniformly iid (independent, identically distributed) sequence.

An iid non-uniform model Let $q_1, q_2, \ldots, q_N \geq 0$ so that

$$\sum_{i=1}^{N} q_i = 1\,.$$

We now construct a probability P on $(\Omega, \mathcal{F})$ so that the $\{X_n\}$ are iid with

$$P(X_n = e_j) = q_j \quad \text{for } j = 1, \ldots, N.$$

Construction At each time n the Markov chain value X_n is just one of the unit vector elements e_i in its state space $\{e_1, e_2, \ldots, e_N\}$.

We shall often use the identity

$$\sum_{j=1}^{N} \langle X_n,\, e_j \rangle = 1$$

for any $n = 0, 1, 2, \ldots$ from time to time without further explanation. Inserting this identity into an argument from time to time is often a useful trick.

For $l = 0, 1, \ldots$, define

$$\overline{\lambda}_l = N\langle q, X_l \rangle$$

where $q = (q_1, \ldots, q_N)^\top$.

Lemma 1.1 *Recall $\overline{E}$ refers to the reference probability $\overline{P}$ defined above.*

(i) $\overline{\mathbf{E}}[\,\overline{\lambda}_0] = 1$,
(ii) $\overline{\mathbf{E}}[\,\overline{\lambda}_l \,|\, \mathcal{F}_{l-1}] = 1$ *for* $l \geq 1$.

Proof For (i), we have

$$\begin{aligned}\overline{\mathbf{E}}[\,\overline{\lambda}_0] = \overline{\mathbf{E}}[N\langle q, X_0\rangle] &= \overline{\mathbf{E}}\left[\sum_{i=1}^{N}\langle X_0, e_i\rangle\, N\langle q, X_0\rangle\right] \\ &= \overline{\mathbf{E}}\left[\sum_{i=1}^{N}\langle X_0, e_i\rangle\, N\,\langle q, e_i\rangle\right] = \sum_{i=1}^{N} N\, q_i \cdot \overline{\mathbf{E}}\left[\langle X_0, e_i\rangle\right] \\ &= \sum_{i=1}^{N} N q_i \cdot \frac{1}{N} = 1.\end{aligned}$$

For (ii), we have

$$\overline{\mathbf{E}}[\,\overline{\lambda}_l|\mathcal{F}_{l-1}] = \overline{\mathbf{E}}[N\langle q, X_l\rangle|\,\mathcal{F}_{l-1}] = \overline{\mathbf{E}}[N\langle q, X_l\rangle] = 1$$

where we used the fact that under $\overline{P}$ the $\{X_k\}$ are independent and so

$$\overline{\mathbf{E}}[N\langle q, X_l\rangle|\,\mathcal{F}_{l-1}] = \overline{\mathbf{E}}[N\langle q, X_l\rangle]$$

and the last equality follows as in $l = 0$. □

We now introduce a new probability on $(\Omega, \mathcal{F})$.
Write

$$\overline{\Lambda}_n = \prod_{l=0}^{n} \overline{\lambda}_n = \overline{\lambda}_0 \cdot \overline{\lambda}_1 \cdots \overline{\lambda}_n\,. \tag{1.2}$$

We define the new probability P by requiring that

$$\left.\frac{dP}{d\overline{P}}\right|_{\mathcal{F}_n} = \overline{\Lambda}_n\,.$$

This simply means that if $A \in \mathcal{F}_n$, then

$$P(A) = \overline{\mathbf{E}}[\overline{\Lambda}_n I(A)]. \tag{1.3}$$

We note that if $A \in \mathcal{F}_n$, then $A \in \mathcal{F}_{n+1}$ also. This leads to two definitions of $P(A)$ depending on whether we use $\overline{\Lambda}_n$ or $\overline{\Lambda}_{n+1}$ in (1.3). However, we have the following result:

Lemma 1.2 *The definition of P is well defined. That is,*

$$\overline{\mathbf{E}}[\overline{\Lambda}_n I(A)] = \overline{\mathbf{E}}[\overline{\Lambda}_m I(A)]$$

for any $A \in \mathcal{F}_n$ and $m > n$.

Proof We first note that

$$\overline{\mathbf{E}}[\overline{\Lambda}_m \mid \mathcal{F}_n] = \overline{\Lambda}_n .$$

This follows from Lemma 1.1, because

$$\begin{aligned}
\overline{\mathbf{E}}[\overline{\Lambda}_m \mid \mathcal{F}_n] &= \overline{\mathbf{E}}[\overline{\mathbf{E}}[\overline{\Lambda}_m \mid \mathcal{F}_{m-1}] \mid \mathcal{F}_n] \\
&= \overline{\mathbf{E}}[\overline{\Lambda}_{m-1}\overline{\mathbf{E}}[\overline{\lambda}_m \mid \mathcal{F}_{m-1}] \mid \mathcal{F}_n] \\
&= \overline{\mathbf{E}}[\overline{\Lambda}_{m-1} \mid \mathcal{F}_n] \\
&= \overline{\mathbf{E}}[\overline{\Lambda}_{m-2} \mid \mathcal{F}_n] \\
&= \cdots \\
&= \overline{\mathbf{E}}[\overline{\Lambda}_n \mid \mathcal{F}_n] \\
&= \overline{\Lambda}_n .
\end{aligned}$$

Then for $A \in \mathcal{F}_n$

$$\begin{aligned}
\overline{\mathbf{E}}[\overline{\Lambda}_m I(A)] &= \overline{\mathbf{E}}[\overline{\mathbf{E}}[\overline{\Lambda}_m I(A) \mid \mathcal{F}_n]] \\
&= \overline{\mathbf{E}}[\overline{\mathbf{E}}[\overline{\Lambda}_m \mid \mathcal{F}_n] I(A)] \\
&= \overline{\mathbf{E}}[\overline{\Lambda}_n I(A)]
\end{aligned}$$

and we are done. □

Now let B be an event in $\mathcal{F}^A$ so $B \in \mathcal{F}_n$ for some $n \geq 0$. We then define

$$P(B) = \overline{\mathbf{E}}[\overline{\Lambda}_n I(B)],$$

and by Lemma 1.2, this is well defined. Suppose $F \in \mathcal{F}$ is of the form

$$F = \bigcup_{j=1}^{\infty} A_j$$

for disjoint events $\{A_j\}$, $A_j \in \mathcal{F}^A$, (we could let many of the $A_j = \emptyset$). We then set

$$P(F) = \sum_{j=1}^{\infty} P(A_j).$$

Properties of the iid non-uniform model We now investigate the statistics of $\{X_k\}$ under P.

Property 1: We have $P(X_k = e_j) = q_j$ for each k, j.

Proof For $k \geq 0$ and $j \in \{1, 2, \ldots, N\}$, let $A = \{\omega \in \Omega | \, \omega_k = e_j\} \in \mathcal{F}_k$, then

$$\begin{aligned}
P(X_k = e_j) &= P\left(\{\omega \in \Omega | \, X_k(\omega) = e_j\}\right) \\
&= P(A) \\
&= \overline{\mathbf{E}}[\overline{\Lambda}_k \, I(A)] \\
&= \overline{\mathbf{E}}\left[\overline{\mathbf{E}}[\overline{\Lambda}_k \, I(A) | \, \mathcal{F}_{k-1}]\right] \\
&= \overline{\mathbf{E}}\left[\overline{\Lambda}_{k-1} \, \overline{\mathbf{E}}[\overline{\lambda}_k \, I(A) | \, \mathcal{F}_{k-1}]\right] \\
&= \overline{\mathbf{E}}\left[\overline{\Lambda}_{k-1} \, \overline{\mathbf{E}}[\overline{\lambda}_k \, I(A)]\right] \\
&= \overline{\mathbf{E}}[\overline{\lambda}_k \, I(A)] \, \overline{\mathbf{E}}[\overline{\Lambda}_{k-1}] \\
&= \overline{\mathbf{E}}[\overline{\lambda}_k \, I(A)],
\end{aligned}$$

where we note that $\overline{\lambda}_k \, I(A)$ depends only on the values of X_k and so is independent under $\overline{P}$ of $\mathcal{F}_{k-1}$.

We also used

$$\overline{\mathbf{E}}[\overline{\Lambda}_{k-1}] = \overline{\mathbf{E}}[\overline{\Lambda}_{k-1} \, I(\Omega)] = P(\Omega) = 1.$$

Continuing the calculation,

$$\begin{aligned}
\overline{\mathbf{E}}[\overline{\lambda}_k \, I(A)] &= \overline{\mathbf{E}}\left[\sum_{i=1}^{N} \langle X_k, e_i \rangle \, N \langle X_k, q \rangle \, I(X_k = e_j)\right] \\
&= \overline{\mathbf{E}}\left[\sum_{i=1}^{N} \langle X_k, e_i \rangle \, N \langle e_i, q \rangle \, I(e_i = e_j)\right] \\
&= \overline{\mathbf{E}}\left[\langle X_k, e_j \rangle \, N q_j\right] \\
&= \frac{1}{N} \cdot N q_j \\
&= q_j
\end{aligned}$$

where we used

$$\overline{\mathbf{E}}[\langle X_k, e_j \rangle] = \overline{P}(X_k = e_j) = \frac{1}{N}.$$

Thus

$$P(X_k = e_j) = q_j$$

as we claimed. $\square$

Property 1 means also that the $\{X_k\}$ have the same distributions for all k.

Property 2: The terms of $\{X_k\}$ are independent random variables.

Proof We show that X_k and X_l are independent for $l > k$ by showing that

$$P(X_k = e_j \,\&\, X_l = e_i) = P(X_k = e_j) \cdot P(X_l = e_i)$$

for any i, j. The proof that we present readily extends to a similar identity for any finite number of terms of the sequence $\{X_k\}$. We leave this to the reader.

We note that

$$\langle X_k, e_j \rangle = \mathrm{I}(X_k = e_j) = \begin{cases} 1 & \text{if } X_k = e_j \\ 0 & \text{if } X_k \neq e_j \,. \end{cases}$$

We have

$$\begin{aligned} &P(X_k = e_j,\, X_l = e_i) \\ &\quad = \overline{\mathbf{E}}\left[\overline{\Lambda}_l\, I(X_k = e_j)\, I(X_l = e_i)\right] \\ &\quad = \overline{\mathbf{E}}\left[\overline{\Lambda}_l\, \langle X_k, e_j \rangle\, \langle X_l, e_i \rangle\right] \\ &\quad = \overline{\mathbf{E}}\left[\overline{\Lambda}_k\, \langle X_k, e_j \rangle\right]\, \overline{\mathbf{E}}\left[\overline{\lambda}_{k+1}, \ldots, \overline{\lambda}_l\, \langle X_l, e_i \rangle\right] \text{ by independence under } \overline{P}, \\ &\quad = q_j\, \overline{\mathbf{E}}\left[\overline{\lambda}_{k+1}, \ldots, \overline{\lambda}_l\, \langle X_l, e_i \rangle\right] \,. \end{aligned}$$

In fact under $\overline{P}$, we saw in the derivation of Property 1, that

$$\overline{\mathbf{E}}\left[\overline{\Lambda}_k\, \langle X_k, e_j \rangle\right] = \overline{\mathbf{E}}\left[\overline{\lambda}_k\, \langle X_k, e_j \rangle\right] = q_j \,.$$

In the same way,

$$\overline{\mathbf{E}}\left[\overline{\lambda}_{k+1}, \ldots, \overline{\lambda}_l\, \langle X_l, e_i \rangle\right] = \overline{\mathbf{E}}\left[\overline{\lambda}_l\, \langle X_l, e_i \rangle\right] = q_i$$

and so we have shown that

$$P(X_k = e_j,\, X_l = e_i) = P(X_k = e_j) \cdot P(X_l = e_i)$$

as claimed. $\square$

1.5 The general Markov chain

The processes we have constructed so far have distributions at each time which are independent of other times. We now construct a Markov chain with a given π as the initial probabilities of X_0 and (a_{ji}) as the transition probabilities. This is again constructed using a change of probability from the reference model.

Write

$$\overline{\lambda}_0 = N \langle \pi, X_0 \rangle$$

where $\pi = (\pi_1, \pi_2, \ldots, \pi_N)^\top$ with $\pi_i \geq 0$ and

$$\sum_{i=1}^{N} \pi_i = 1 .$$

With $A = (a_{ji})$ for $l \geq 1$, set

$$\overline{\lambda}_l = N\langle X_l,\, AX_{l-1}\rangle .$$

Recall $\sum_{j=1}^{N} a_{ji} = 1$.

Lemma 1.3 *We have the properties:*

(i) $\overline{\mathbf{E}}[\overline{\lambda}_0] = 1$;
(ii) $\overline{\mathbf{E}}[\overline{\lambda}_l | \mathcal{F}_{l-1}] = 1$.

Proof Part (i) is shown as in Lemma 1.1. For (ii)

$$\begin{aligned}
\overline{\mathbf{E}}[\overline{\lambda}_l | \mathcal{F}_{l-1}] &= \overline{\mathbf{E}}[N\langle X_l,\, AX_{l-1}\rangle | \mathcal{F}_{l-1}] \\
&= \overline{\mathbf{E}}\left[\sum_{j=1}^{N} \langle X_l,\, e_j\rangle\, N\langle X_l,\, AX_{l-1}\rangle | \mathcal{F}_{l-1}\right] \\
&= \overline{\mathbf{E}}\left[\sum_{j=1}^{N} \langle X_l,\, e_j\rangle\, N\langle e_j,\, AX_{l-1}\rangle | \mathcal{F}_{l-1}\right] \\
&= \sum_{j=1}^{N} N\langle e_j,\, AX_{l-1}\rangle\, \overline{\mathbf{E}}\left[\, \langle X_l,\, e_j\rangle | \mathcal{F}_{l-1}\right] \\
&= \sum_{j=1}^{N} \langle e_j,\, AX_{l-1}\rangle
\end{aligned}$$

as $\overline{\mathbf{E}}\left[\, \langle X_l,\, e_j\rangle | \mathcal{F}_{l-1}\right] = \overline{\mathbf{E}}\left[\, \langle X_l,\, e_j\rangle\right] = \overline{P}(X_l = e_j) = 1/N$.

So

$$\overline{\mathbf{E}}[\overline{\lambda}_l | \mathcal{F}_{l-1}] = \sum_{j=1}^{N} \sum_{i=1}^{N} \langle X_{l-1}, e_i \rangle \langle e_j, AX_{l-1} \rangle$$
$$= \sum_{j=1}^{N} \sum_{i=1}^{N} \langle X_{l-1}, e_i \rangle \langle e_j, A e_i \rangle$$
$$= \sum_{j=1}^{N} \sum_{i=1}^{N} a_{ji} \langle X_{l-1}, e_i \rangle = \sum_{i=1}^{N} \langle X_{l-1}, e_i \rangle = 1,$$

as $\sum_{j=1}^{N} a_{ji} = 1$. □

Now write

$$\overline{\Lambda}_n = \prod_{l=0}^{N} \overline{\lambda}_l$$

and define P on $(\Omega, \mathcal{F})$ by

$$\left.\frac{dP}{d\overline{P}}\right|_{\mathcal{F}_n} = \overline{\Lambda}_n.$$

This leads to P being well defined on $\mathcal{F}$ in the way described before.

Properties of this model

Property 1: We have $P(X_0 = e_j) = \pi_j$ for $j = 1, 2, \ldots, N$.
This is proved just as in the second model.

Property 2: We have $P(X_{k+1} = e_j | X_k = e_i) = a_{ji}$ for all i, j and all k.

Proof We use

$$P(X_{k+1} = e_j \,|\, X_k = e_i) = \frac{P(X_{k+1} = e_j \,,\, X_k = e_i)}{P(X_k = e_i)}.$$

We have

$$P(X_{k+1} = e_j \,,\, X_k = e_i) = \overline{\mathbf{E}}\left[\overline{\Lambda}_{k+1} \langle X_{k+1}, e_j \rangle \langle X_k, e_i \rangle\right]$$
$$= \overline{\mathbf{E}}\left[\overline{\mathbf{E}}[\overline{\Lambda}_{k+1} \langle X_{k+1}, e_j \rangle \langle X_k, e_i \rangle | \mathcal{F}_k]\right]$$
$$= \overline{\mathbf{E}}\left[\overline{\Lambda}_k \langle X_k, e_i \rangle \overline{\mathbf{E}}[\overline{\lambda}_{k+1} \langle X_{k+1}, e_j \rangle | \mathcal{F}_k]\right].$$

Now

$$
\begin{aligned}
&\overline{\mathbf{E}}[\,\overline{\lambda}_{k+1}\,\langle X_{k+1}, e_j\rangle |\, \mathcal{F}_k] \\
&\quad = \overline{\mathbf{E}}\,[\,N\,\langle X_{k+1}, AX_k\rangle\,\langle X_{k+1}, e_j\rangle |\, \mathcal{F}_k] \\
&\quad = \overline{\mathbf{E}}\left[\sum_{r,\,s=1}^{N} \langle X_{k+1}, e_r\rangle\,\langle X_k, e_s\rangle\, N\langle X_{k+1}, AX_k\rangle\,\langle X_{k+1}, e_j\rangle |\, \mathcal{F}_k\right] \\
&\quad = \overline{\mathbf{E}}\left[\sum_{r,\,s=1}^{N} \langle X_{k+1}, e_r\rangle\langle X_k, e_s\rangle\, N\cdot a_{rs}\,\delta_{rj} |\, \mathcal{F}_k\right] \\
&\quad = \overline{\mathbf{E}}\left[\sum_{s=1}^{N}\langle X_{k+1}, e_j\rangle\langle X_k, e_s\rangle\, N\cdot a_{js} |\, \mathcal{F}_k\right] \\
&\quad = \sum_{s=1}^{N} a_{js}\,\langle X_k, e_s\rangle\, N\,\overline{\mathbf{E}}[\langle X_{k+1}, e_j\rangle |\, \mathcal{F}_k] \\
&\quad = \sum_{s=1}^{N} a_{js}\,\langle X_k, e_s\rangle
\end{aligned}
$$

as

$$\overline{\mathbf{E}}[\langle X_{k+1}, e_j\rangle |\, \mathcal{F}_k] = \overline{\mathbf{E}}[\langle X_{k+1}, e_j\rangle] = \frac{1}{N}$$

as above.

Putting these calculations together,

$$
\begin{aligned}
P(X_{k+1} = e_j\,,\, X_k = e_i) &= \overline{\mathbf{E}}\left[\overline{\Lambda}_k\langle X_k, e_i\rangle \sum_{s=1}^{N} a_{js}\langle X_k, e_s\rangle\right] \\
&= \overline{\mathbf{E}}\left[\overline{\Lambda}_k\langle X_k, e_i\rangle \sum_{s=1}^{N} a_{js}\langle e_i, e_s\rangle\right] \\
&= \overline{\mathbf{E}}\left[\overline{\Lambda}_k\langle X_k, e_i\rangle \sum_{s=1}^{N} a_{js}\,\delta_{is}\right] \\
&= a_{ji}\,\overline{\mathbf{E}}[\overline{\Lambda}_k\langle X_k, e_i\rangle] \\
&= a_{ji}\,P(X_k = e_i)
\end{aligned}
$$

The proof is complete. □

As before write

$$A = (a_{ji})$$

where

$$a_{ji} = P(X_{n+1} = e_j | X_n = e_i).$$

A semi-martingale representation of $\{\mathbf{X_n}\}$ The semi-martingale representation of the vector-valued process X_{n+1} is its decomposition into the sum of an $\mathcal{F}_n$-measurable term and a vector-valued martingale increment. If we define

$$V_{n+1} = X_{n+1} - AX_n \,.$$

Then clearly

$$X_{n+1} = AX_n + V_{n+1} \,. \tag{1.4}$$

Certainly AX_n is $\mathcal{F}_n$-measurable. We now show that

$$\mathbf{E}[V_{n+1}|\,\mathcal{F}_n] = \mathbf{E}[V_{n+1}|\,X_n] = 0 \in \mathbb{R}^N.$$

This will establish the following result.

Theorem 1.4 *The semi-martingale representation of* $\{X_n\}$ *is* $X_{n+1} = AX_n + V_{n+1}$.

Proof Taking conditional expectations, we have

$$\mathbf{E}[V_{n+1}\,|\,X_n] = \mathbf{E}[X_{n+1} - AX_n\,|\,X_n] = \mathbf{E}[X_{n+1}\,|\,X_n] - AX_n \,.$$

We now compute $\mathbf{E}[X_{n+1}\,|\,X_n]$.

Expressing X_{n+1} as

$$X_{n+1} = \sum_j \langle X_{n+1}, e_j\rangle\, e_j = \sum_j \mathrm{I}(X_{n+1} = e_j)\, e_j$$

we have for any i,

$$\begin{aligned}
\mathbf{E}[X_{n+1}\,|\,X_n = e_i] &= \mathbf{E}\left[\sum_j \mathrm{I}(X_{n+1} = e_j)\, e_j \,\middle|\, X_n = e_i\right] \\
&= \sum_j e_j\, \mathbf{E}\left[\mathrm{I}(X_{n+1} = e_j)\,|\,X_n = e_i\right] \\
&= \sum_j e_j\, \mathrm{P}(X_{n+1} = e_j\,|\,X_n = e_i) \\
&= \sum_j e_j\, a_{ji} \\
&= A\, e_i
\end{aligned}$$

and so $\mathbf{E}[X_{n+1}\,|\,X_n] = AX_n$. Consequently, with $V_{n+1} = X_{n+1} - AX_n$

$$E[V_{n+1}|X_n] = E[X_{n+1} - AX_n|X_n] = 0 \in \mathbb{R}^N,$$

so V_{n+1} is a vector martingale increment. The proof is complete. □

We can also calculate the conditional variance–covariance matrix

$$\mathbf{E}[V_{n+1}\, V_{n+1}^{\top} \,|\, X_n]\,.$$

Here, V_{n+1} is $N \times 1$ vector and $V_{n+1}\, V_{n+1}'$ is an $N \times N$ matrix.

The (i,j) element of $\mathbf{E}[V_{n+1}\, V_{n+1}^{\top} \,|\, X_n]$ can be written as

$$\mathbf{E}[V_{n+1,i}\, V_{n+1,j} \,|\, X_n].$$

The evaluation of this conditional variance–covariance matrix is given in the following lemma.

Lemma 1.5 *We have*

$$\mathbf{E}[V_{n+1}\, V_{n+1}^{\top} \,|\, X_n] = \operatorname{diag}(AX_n) - A \operatorname{diag}(X_n)\, A^{\top}\,.$$

Proof We note that

$$X_{n+1}\, X_{n+1}^{\top} = \operatorname{diag}(X_{n+1})\,.$$

For example,

$$\begin{bmatrix} 1 \\ 0 \end{bmatrix} \begin{bmatrix} 1 & 0 \end{bmatrix} = \begin{bmatrix} 1 & 0 \\ 0 & 0 \end{bmatrix} = \operatorname{diag}\left[\begin{bmatrix} 1 \\ 0 \end{bmatrix}\right].$$

This implies that

$$X_{n+1}\, X_{n+1}^{\top} = \operatorname{diag}(AX_n) + \operatorname{diag}(V_{n+1})\,.$$

and hence

$$\mathbf{E}[X_{n+1}\, X_{n+1}^{\top} \,|X_n] = \operatorname{diag}(AX_n).$$

We can also write

$$\begin{aligned} X_{n+1}\, X_{n+1}^{\top} &= (AX_n + V_{n+1})(AX_n + V_{n+1})^{\top} \\ &= (AX_n + V_{n+1})(X^{\top} A_n^{\top} + V_{n+1}^{\top}) \\ &= AX_n\, X_n^{\top} A^{\top} + V_{n+1} X_n^{\top} A^{\top} + AX_n V_{n+1}^{\top} + V_{n+1}\, V_{n+1}^{\top} \\ &= A \operatorname{diag}(X_n)\, A^{\top} + V_{n+1} X_n^{\top} A^{\top} + AX_n V_{n+1}^{\top} + V_{n+1}\, V_{n+1}^{\top} \end{aligned}$$

and so, as V_{n+1} is a martingale increment,

$$\mathbf{E}[X_{n+1}\, X_{n+1}^{\top} | X_n] = A \operatorname{diag}(X_n)\, A^{\top} + \mathbf{E}[V_{n+1}\, V_{n+1}^{\top} \,|X_n].$$

Comparing the two expressions, we obtain

$$\mathbf{E}[V_{n+1}\, V_{n+1}^{\top} \,|X_n] = \operatorname{diag}(AX_n) - A \operatorname{diag}(X_n)\, A^{\top}$$

and so the lemma is proved. □

Remark These arguments apply to any first-order Markov chain.

The decomposition (1.4) is sometimes called a semi-martingale decomposition. This is because X_{n+1} is expressed as a predictable part, AX_n, and a noise term, or martingale increment term, which has conditional mean zero. The decomposition in (1.4) is unique.

Strictly speaking a semi-martingale representation of $\{X_n\}$ means, in this context, that for each n

$$X_n = \mathcal{A}_n + \mathcal{M}_n$$

where $\{\mathcal{A}_n\}$ is a predictable process and $\{\mathcal{M}_n\}$ is a $\{\mathcal{F}_n\}$-martingale. If fact, this decomposition is unique and for

$$\mathcal{A}_n = X_0 + \sum_{k=1}^{n+1} (A - \mathrm{I})\ X_{k-1} \quad \text{and} \quad \mathcal{M}_n = \sum_{k=1}^{n+1} V_k\,. \tag{1.5}$$

The representations (1.4) and (1.5) are equivalent.

1.6 Conclusion

In this opening chapter Markov chains have been constructed from first principles. The state space is the set of all sequences whose elements are points in the state space. The probability measures are constructed explicitly so that the terms in the sequences are independent random variables or finally Markov chains with given transition probabilities. These constructions are not usually given in the literature.

1.7 Exercises

Exercise 1.1 Explicitly construct a Markov chain on the state space $\{e_1, e_2, e_3\}$ for $t \in \{0, 1, 2, 3\}$ with transition matrix

$$\begin{pmatrix} 1/2 & 1/3 & 0 \\ 1/4 & 1.3 & 1/2 \\ 1/4 & 1/3 & 1/2 \end{pmatrix}.$$

Exercise 1.2 Prove the decomposition (1.4) is unique.

2

Estimation of an Observed Markov Chain

2.1 Introduction

The first chapter constructed Markov chains with prescribed transition probabilities. This chapter discusses how, by observing sample paths of a Markov chain, those probabilities can be estimated. The probability distribution of the initial state is also estimated. The proofs are based on maximum likelihood estimation where the likelihoods are just the densities use in Chapter 1. Chains of higher order are also mentioned.

We now discuss how to estimate the parameters of a Markov chain. Recall that a Markov chain is specified by giving $\pi = (\pi_j)$, the initial probabilities, and the matrix $A = (a_{ji})$ of transition probabilities.

Suppose we observe the Markov chain. We wish to estimate $A = (a_{ji})$ and $\pi = (\pi_j)$.

(a) In economics, finance, or tracking, there is only one sample path

$$(x_0, x_1, x_2, x_3, \ldots, x_L).$$

In this case, we can estimate (a_{ji}) but not (π_j). On the other hand, in these applications, the effect of x_0 often dies away so the initial value is not important. (The Kalman filtering literature has extensive discussion of these 'forgetting' properties.)

(b) In genomics, we have many sample paths of a chain. Then it is possible to estimate (π_j). See Durbin et al. (1998).

2.2 Estimation based on one sample path

In the construction of Section 1.4 the Λ defined in (1.2) is the density or likelihood. Given a sequence of observations we shall determine the transition probabilities which maximize this likelihood.

Note that, with $\overline{\Lambda}_L$ defined as in (1.2),

$$\begin{aligned}
&P(X_0 = x_0, X_1 = x_1, \ldots, X_L = x_l \mid \pi, A) \\
&= \overline{\mathbf{E}}\left[\overline{\Lambda}_L \, \mathrm{I}(X_0 = x_0, X_1 = x_1, \ldots, X_L = x_l)\right] \\
&= \overline{\mathbf{E}}\left[\prod_{l=0}^{L} \overline{\lambda}_l \, \mathrm{I}(X_0 = x_0, X_1 = x_1, \ldots, X_L = x_l)\right] \\
&= \overline{\mathbf{E}}\left[N\langle \pi, X_0\rangle \prod_{l=1}^{L} N\langle X_l, A\,X_{l-1}\rangle \, \mathrm{I}(X_0 = x_0, \ldots, X_L = x_l)\right] \\
&= \frac{1}{N^{L+1}} N\langle \pi, x_0\rangle \prod_{l=1}^{L} N\langle x_l, A\,x_{l-1}\rangle = \langle \pi, x_0\rangle \prod_{l=1}^{L} \langle x_l, A\,x_{l-1}\rangle
\end{aligned}$$

$\equiv \mathrm{L}(\pi, A \mid x_0, \ldots, x_L)$, say.

Write

$$\begin{aligned}
\ell(\pi, A \mid x_0, \ldots, x_L) &= \log \mathrm{L}(\pi, A \mid x_0, \ldots, x_L) \\
&= \log \langle \pi, x_0\rangle + \sum_{l=1}^{L} \log \langle x_l, A\,x_{l-1}\rangle\,.
\end{aligned}$$

This will be maximized with respect to A, (and with respect to π later), under the constraints

$$a_{ji} \geq 0, \qquad \sum_{j=1}^{N} a_{ji} = 1\,.$$

As all the $x_l \in \mathcal{S} = \{e_1, e_2, \ldots, e_N\}$ and $\sum_{r=1}^{N} \langle x_l, e_r\rangle = 1$ we can write

$$\begin{aligned}
&\ell(\pi, A \mid x_0, \ldots, x_L) \\
&= \log \langle \pi, x_0\rangle + \sum_{l=1}^{L} \sum_{r,s=1}^{N} \langle x_l, e_r\rangle \langle x_{l-1}, e_s\rangle \log \langle x_l, A\,x_{l-1}\rangle \\
&= \log \langle \pi, x_0\rangle + \sum_{l=1}^{L} \sum_{r,s=1}^{N} \langle x_l, e_r\rangle \langle x_{l-1}, e_s\rangle \log a_{rs}\,. \qquad (2.1)
\end{aligned}$$

We consider a fixed s, $1 \leq s \leq N$, and maximize $\ell(\pi, A \mid x_0, \ldots, x_L)$ under the constraints

$$a_{rs} \geq 0, \qquad \sum_{r=1}^{N} a_{rs} = 1\,.$$

Notation 2.1 *For any vector* $q = (q_1, q_2, \ldots, q_N)^\top \in \mathbb{R}^N$ *we write* $q \gg 0$ *if* $q_i > 0$ *for all* i, $1 \le i \le N$.

Lemma 2.2 *If* $q \gg 0$, $\hat{q} \gg 0$ *with* $\sum_{i=1}^N q_i = 1$, $\sum_{i=1}^N \hat{q}_i = 1$, *and* α_1, $\alpha_2, \ldots, \alpha_N > 0$, *then*

$$\sum_{i=1}^N \alpha_i \log q_i \le \sum_{i=1}^N \alpha_i \log \hat{q}_i + \sum_{i=1}^N \alpha_i \frac{1}{\hat{q}_i} (q_i - \hat{q}_i) .$$

Proof Suppose $0 < x < y$. Then

$$\begin{aligned} \log x - \log y &= \int_x^y -\frac{1}{t}\, dt \\ &\le -\frac{1}{y} \int_x^y dt \\ &= -\frac{1}{y}(y - x) \\ &= \frac{1}{y}(x - y) . \end{aligned}$$

If $x > y > 0$, then

$$\begin{aligned} \log x - \log y &= \int_y^x \frac{1}{t}\, dt \\ &\le \frac{1}{y} \int_y^x dt \\ &= \frac{1}{y}(x - y) . \end{aligned}$$

Thus, $\log x \le \log y + \frac{1}{y}(x - y)$ for any $x, y > 0$. This proves the lemma. □

Lemma 2.3 *Let* $\alpha_1, \alpha_2, \ldots, \alpha_N \ge 0$ *with* $\alpha_1 + \cdots + \alpha_N > 0$. *Then the objective*

$$I = I(q) = \sum_{i=1}^N \alpha_i \log q_i$$

has a maximum over the set

$$\mathcal{U} = \left\{ q = (q_1, \ldots, q_N) \mid q_i \ge 0 \ \ \textit{for all } i,\ \sum_{i=1}^N q_i = 1 \right\}$$

and the maximum is attained at q so that

$$q_i = \hat{q}_i = \alpha_i \left(\sum_{i=1}^{N} \alpha_i \right)^{-1} \quad \text{for all } i = 1, 2, \dots, N.$$

Proof Note that the maximum of $I(q)$ over $\mathcal{U}$ is the same as the maximum of $I(q)$ over $\mathcal{U}'$, a subset of $\mathcal{U}$ given by

$$\mathcal{U}' = \left\{ q \in \mathcal{U} \;\middle|\; I(q) \geq \log\left(\frac{1}{N}\right) \sum_{i=1}^{N} \alpha_i \right\},$$

since the maximum of $I(q)$ is as least as large as the value at of I at the q where $q_i = 1/N$ for each i.

The set $\mathcal{U}'$ is a compact (bounded and closed) subset of $\mathbb{R}^N$. The function I is continuous on $\mathcal{U}'$ and so achieves an upper bound on this set at some point $q = \hat{q}$. This maximizing point is unique, as I is strictly concave on $\mathcal{U}'$, and $I(\hat{q}) \leq 0$. As each $\alpha_i > 0$, it follows that $\hat{q}_i > 0$ for each i.

Now if $q \in \mathcal{U}'$, by the previous lemma, we have

$$\sum_{i=1}^{N} \alpha_i \log q_i \leq \sum_{i=1}^{N} \alpha_i \log \hat{q}_i + \sum_{i=1}^{N} \alpha_i \frac{1}{\hat{q}_i} (q_i - \hat{q}_i).$$

If $\hat{q}_i = \lambda \alpha_i$, then because $\sum_{i=1}^{N} q_i = \sum_{i=1}^{N} \hat{q}_i = 1$,

$$\sum_{i=1}^{N} \alpha_i \frac{1}{\hat{q}_i} (q_i - \hat{q}_i) = 0.$$

Therefore, for $\hat{q}_i = \lambda \alpha_i$ we have that $I(q) \leq I(\hat{q})$. So q defined by

$$\hat{q}_i = \alpha_i \left(\sum_{i=1}^{N} \alpha_i \right)^{-1}$$

maximizes $I(q)$ over $\mathcal{U}'$ and we are done. □

An application Returning to (2.1) for each fixed $s = 1, 2, \dots, N$, we wish to maximize

$$I(a_s) = \sum_{r=1}^{N} \left(\sum_{l=1}^{L} \langle x_l, e_r \rangle \langle x_{l-1}, e_s \rangle \right) \log a_{rs}$$

where $a_s = (a_{1s}, a_{2s}, \dots, a_{Ns})^\top$ is subject to

$$a_{rs} \geq 0 \text{ for each } r \text{ and } \sum_{r=1}^{N} a_{rs} = 1.$$

Lemma 2.3 provides the optimizer

$$\hat{a}_{rs} = \frac{\sum_{l=1}^{L} \langle x_l, e_r \rangle \langle x_{l-1}, e_s \rangle}{\sum_{l=1}^{L} \sum_{r'=1}^{N} \langle x_l, e_{r'} \rangle \langle x_{l-1}, e_s \rangle} = \frac{\sum_{l=1}^{L} \langle x_l, e_r \rangle \langle x_{l-1}, e_s \rangle}{\sum_{l=1}^{L} \langle x_{l-1}, e_s \rangle} = \frac{J_L^{rs}}{O_L^s}$$

where we used

$$\sum_{r'=1}^{N} \langle x_l, e_{r'} \rangle = 1$$

and introduced the notations

$$\begin{aligned} J_L^{rs} &= \text{number of transitions } e_s \text{to } e_r \text{ in the sequence } (x_0, \dots, x_L), \\ O_L^s &= \text{number of times } x_l = e_s \text{ in the sequence } (x_0, \dots, x_{L-1}). \end{aligned}$$

We have above:

$$\hat{a}_{rs} = \frac{\sum_{l=1}^{L} \langle x_l, e_r \rangle \langle x_{l-1}, e_s \rangle}{\sum_{l=1}^{L} \langle x_{l-1}, e_s \rangle} = \frac{J_L^{rs}}{O_L^s}$$

In the case of sparse data, the denominator could be zero or close to zero. In that case, additive smoothing or Laplace smoothing can be used (see Chen and Goodman, 1996). This means we replace the above expression with

$$\hat{a}_{rs} = \frac{\sum_{l=1}^{L} \langle x_l, e_r \rangle \langle x_{l-1}, e_s \rangle + \alpha}{\sum_{l=1}^{L} \langle x_{l-1}, e_s \rangle + N\,\alpha}$$

where $\alpha > 0$. It is still the case that $\hat{a}_{rs} > 0$ and

$$\sum_{s=1}^{N} \hat{a}_{rs} = 1$$

for each r. A review of techniques like this is given in Chen and Goodman (1996), who in turn cite Jelinek and Mercer (1980), and discuss choices for α.

Remark These proofs already show some of the advantages of why the state space $\mathcal{S}$ was chosen. The proof of the estimator here is adapted from (Elliott et al., 1995, page 36ff.). Other arguments have been adapted from Aggoun and Elliott (2004).

2.3 Estimation using K sample paths of length L

If we were to select π based on one sample path, then we would select the MLE choice as $\hat{\pi}_i = 1$ if $X_0 = e_i$ and $\hat{\pi}_j = 0$ for $j \neq i$. This would not be very satisfactory. However if we have a number of sample paths, then we can make more meaningful estimates for π.

Suppose we now have K iid chains

$$\{X_k^j : k = 0, 1, \ldots\},\ j = 1, \ldots, K.$$

Suppose we observe K sample paths of length L:

$$\{x_k^j : k = 0, 1, \ldots, L\},\ j = 1, \ldots, K.$$

We give some details when $K = 2$. The sample space Ω^2 now consists of all sequences of the form:

$$\omega = (\omega_0, \omega_1, \omega_2, \ldots)$$

where $\omega_i \in \mathcal{S}^2$ for each $i \geq 0$. This means that

$$\omega_k = \begin{pmatrix} \omega_k^1 \\ \omega_k^2 \end{pmatrix}$$

where $\omega_k^1, \omega_k^2 \in \mathcal{S}$ for each k. Consider the family $\mathcal{F}^{2A}$ of subsets of Ω of the form $\{\omega \in \Omega^2 \mid \omega_{i_k} = (e_{i_k}^1, e_{i_k}^2),\quad k = 1, 2, \ldots, e\}$. Then $\mathcal{F}^2$ is the σ-algebra generated by $\mathcal{F}^{2A}$.

Each element of $\mathcal{F}^2$ can be expressed as an (at most countable) union of sets in $\mathcal{F}^{2A}$ to which we assign probabilities

$$\mathbb{P}(A) = P(A^1) \times P(A^2)$$

where

$$A_1 = \left\{\omega \in \Omega \mid \omega_{i_k}^1 = e_{i_k^1}, k = 1, 2, \ldots, l\right\}$$

$$A_2 = \left\{\omega \in \Omega \mid \omega_{i_j}^2 = e_{i_j^1}, k = 1, 2, \ldots, l\right\}.$$

Here $(\Omega, \mathcal{F}, P)$ is the probability space defined in Chapter 1. The same probability function P is applied to A_1 and A_2 above, which are projections of $A \in \mathcal{F}^2$ into two elements $A_1, A_2 \in \mathcal{F}$.

The two canonical processes are $\{X_k^i\}$ for $i = 1, 2$ defined on Ω by

$$X_k^i(\omega) = \omega_k^i \quad \text{for each } \omega \in \Omega$$

for $k = 0, 1, 2, \ldots$

Because of the way we defined the probability $\mathbb{P}$ on $(\Omega^2, \mathcal{F}^2)$, these two processes are independent and identically distributed for any choice of probability P on $(\Omega, \mathcal{F})$. This construction easily generalizes to $K > 2$.

In fact suppose $K \geq 2$. We note that for $j = 1, 2$:

$$P(X_0^j = x_0^j, \ldots, X_L^j = x_L^j \mid \pi, A) = \langle \pi, x_0^j \rangle \prod_{l=1}^{L} \langle x_l^j, A\, x_{l-1}^j \rangle\,.$$

For each $k \in \{0, 1, \ldots, L\}$, let

$$\mathbf{X}_k = \left(X_k^1, \ldots, X_k^K\right)^\top \quad \text{with} \quad X_k^j \in \mathcal{S}$$

and suppose we observed a sequence of values of these K chains:

$$\mathbf{x}_k = \left(x_k^1, \ldots, x_k^K\right)^\top.$$

Then, for these K observed chains,

$$\begin{aligned} &P(\mathbf{X}_0 = \mathbf{x}_0, \mathbf{X}_1 = \mathbf{x}_1, \ldots, \mathbf{X}_L = \mathbf{x}_L \mid \pi, A) \\ &= \prod_{j=1}^{K} \left(\langle \pi, x_0^j \rangle \prod_{l=1}^{L} \langle x_l^j, A\, x_{l-1}^j \rangle \right). \end{aligned}$$

Let $\mathbf{X} = (\mathbf{X}_0, \ldots, \mathbf{X}_L)$ and $\mathbf{x} = (\mathbf{x}_0, \ldots, \mathbf{x}_L)$. The log-likelihood function here is

$$\begin{aligned} &L(\pi, A \mid \mathbf{X} = \mathbf{x}) \\ &= \sum_{j=1}^{K} \log \langle \pi, x_0^j \rangle + \sum_{j=1}^{K} \sum_{l=1}^{L} \sum_{r,s=1}^{N} \langle x_l^j, e_r \rangle \langle x_{l-1}^j, e_s \rangle \log a_{rs} \\ &= \sum_{j=1}^{K} \sum_{r=1}^{N} \langle x_0^j, e_r \rangle \log \pi_r + \sum_{j=1}^{K} \sum_{l=1}^{L} \sum_{r,s=1}^{N} \langle x_l^j, e_r \rangle \langle x_{l-1}^j, e_s \rangle \log a_{rs}\,. \end{aligned}$$

This, slightly different, likelihood function is maximized with the following choices (estimators)

$$\begin{aligned} \hat{\pi}_r &= \frac{\sum_{j=1}^{K} \langle x_0^j, e_r \rangle}{\sum_{j=1}^{K} \sum_{r'=1}^{N} \langle x_0^j, e_{r'} \rangle} = \frac{1}{K} \sum_{j=1}^{K} \langle x_0^j, e_r \rangle \\ &= \frac{\#\left\{ j \in \{1, 2, \ldots, K\} \text{ with } x_0^j = e_r \right\}}{K} \end{aligned}$$

and

$$\hat{a}_{rs} = \frac{\sum_{j=1}^{K} \sum_{l=1}^{L} \langle x_l^j, e_r \rangle \langle x_{l-1}^j, e_s \rangle}{\sum_{j=1}^{K} \sum_{l=1}^{L} \langle x_{l-1}^j, e_s \rangle} = \frac{J_{KL}^{rs}}{O_{KL}^{s}}\,.$$

Remark Can $O_L^r = 0$? That is, we never observe the chain in state e_r.

Note that

$$O_L^r = \sum_{l=1}^{L} \langle x_{l-1}, e_r \rangle = 0$$

when $x_0 \neq e_r, x_1 \neq e_r, \ldots, x_{L-1} \neq e_r$. This occurs with probability

$$\overline{P} = \left(1 - \frac{1}{N}\right)^L .$$

This quantity is small if L is large enough. For example, if $N = 4$ and $L = 100$ the answer is $3.2072 \times 10^{-13} \approx 0$.

2.4 Markov chains of order $M \geq 2$

We here give some initial results on chains of higher order. More details are given in Appendices A and B.

The case $M = N = 2$. As before, the state space is $\mathcal{S} = \{e_1, e_2, \ldots, e_N\}$. For a chain of order 2:

$$P(X_{n+1} = e_k \,|\, \mathcal{F}_n) = P[X_{n+1} = e_k \,|\, X_n, X_{n-1}].$$

We can define π, the initial probabilities. However, now:

$$a_{k,ji} = P[X_{n+1} = e_k \,|\, X_n = e_j, X_{n-1} = e_i].$$

Our goal is to re-express this order-2 Markov chain as an order-1 Markov chain, and then apply the previous results. We would then like to generalize these results to order $M > 2$.

Consider the dynamics of $(X_n, X_{n-1}) \to (X_{n+1}, X_n)$.

Suppose $X_n = e_j$ and $X_{n-1} = e_i$. Consider the Kronecker (or direct) product $X_n \otimes X_{n-1}$. For example, if $\mathcal{S} = \{e_1, e_2\}$ with e_1, e_2 the standard basis vectors in $\mathbb{R}^2$:

$$e_1 \otimes e_2 = \begin{bmatrix} 1 \\ 0 \end{bmatrix} \otimes \begin{bmatrix} 0 \\ 1 \end{bmatrix} = \begin{bmatrix} 1 \begin{bmatrix} 0 \\ 1 \end{bmatrix} \\ \\ 0 \begin{bmatrix} 0 \\ 1 \end{bmatrix} \end{bmatrix} = \begin{bmatrix} 0 \\ 1 \\ 0 \\ 0 \end{bmatrix}.$$

In general if matrix A is of size $m \times n$ and matrix B is of size $p \times q$ then we define the $mp \times nq$ matrix

$$A \otimes B = \begin{bmatrix} a_{11}B & a_{12}B & \cdots & a_{1n}B \\ \vdots & \vdots & \ddots & \vdots \\ a_{m1}B & a_{m2}B & \cdots & a_{mn}B \end{bmatrix}.$$

Taking $e_1, e_2 \in \mathbb{R}^n$ and the Kronecker products in lexicographical (dictionary order), we have

$$e_1 \otimes e_1 = \begin{bmatrix} 1 \\ 0 \\ 0 \\ 0 \end{bmatrix} = f_1$$

$$e_1 \otimes e_2 = \begin{bmatrix} 0 \\ 1 \\ 0 \\ 0 \end{bmatrix} = f_2$$

$$e_2 \otimes e_1 = \begin{bmatrix} 0 \\ 0 \\ 1 \\ 0 \end{bmatrix} = f_3$$

$$e_2 \otimes e_2 = \begin{bmatrix} 0 \\ 0 \\ 0 \\ 1 \end{bmatrix} = f_4\,,$$

where f_1, f_2, f_3, f_4 are the standard unit vectors in $\mathbb{R}^4$. Conversely, given $f_k = e_i \otimes e_i$, we can find the e_i and e_j using the following identities:

$$\begin{bmatrix} 1 & 1 & 0 & 0 \\ 0 & 0 & 1 & 1 \end{bmatrix} e_i \otimes e_j = e_i\,,$$

$$\begin{bmatrix} 1 & 0 & 1 & 0 \\ 0 & 1 & 0 & 1 \end{bmatrix} e_i \otimes e_j = e_j\,.$$

If we define

$$J_1 = \begin{bmatrix} 1 & 1 & 0 & 0 \\ 0 & 0 & 1 & 1 \end{bmatrix} \quad \text{and} \quad J_2 = \begin{bmatrix} 1 & 0 & 1 & 0 \\ 0 & 1 & 0 & 1 \end{bmatrix}.$$

Then

$$J_1 \left(X_n \otimes X_{n-1} \right) = X_n$$

and

$$J_2 (X_n \otimes X_{n-1}) = X_{n-1} .$$

There is thus a one-to-one correspondence between (X_n, X_{n-1}) and $X_n \otimes X_{n-1}$. Note that in the transitions $(X_n, X_{n-1}) \to (X_{n+1}, X_n)$ not every transition $e_i \otimes e_j \to e_r \otimes e_s$ is possible. We must always have $i = s$. This explains the 0's in the next table of transition probabilities which gives the transition matrix Π for the order 2 chain:

$X_{n+1} \otimes X_n$	Π				$X_n \otimes X_{n-1}$
$e_1 \otimes e_1$	$a_{1,11}$	$a_{1,12}$	0	0	$e_1 \otimes e_1$
$e_1 \otimes e_2$	0	0	$a_{1,21}$	$a_{1,22}$	$e_1 \otimes e_2$
$e_2 \otimes e_1$	$a_{2,11}$	$a_{2,12}$	0	0	$e_2 \otimes e_1$
$e_2 \otimes e_2$	0	0	$a_{2,21}$	$a_{2,22}$	$e_2 \otimes e_2$

Define $Z_n = X_n \otimes X_{n-1}$. Then $\mathrm{P}(Z_{n+1} = f_j \mid Z_n = f_i)$ is specified by the 4×4 matrix Π, as above, with

$$f_1 = \begin{bmatrix} 1 \\ 0 \\ 0 \\ 0 \end{bmatrix}, \quad f_2 = \begin{bmatrix} 0 \\ 1 \\ 0 \\ 0 \end{bmatrix}, \quad f_3 = \begin{bmatrix} 0 \\ 0 \\ 1 \\ 0 \end{bmatrix}, \quad f_4 = \begin{bmatrix} 0 \\ 0 \\ 0 \\ 1 \end{bmatrix}.$$

Remark We would like to find a convenient way of expressing Π in terms of

$$A = \begin{bmatrix} a_{1,11} & \cdots & a_{1,22} \\ a_{2,11} & \cdots & a_{2,22} \end{bmatrix}.$$

The converse is much easier :

$$A = J_1 \, \Pi = \begin{bmatrix} 1 & 1 & 0 & 0 \\ 0 & 0 & 1 & 1 \end{bmatrix} \Pi.$$

We shall generalize these results to the case when $M > 2$ and $N \geq 2$ in Appendix 1.

Remark There are various approaches to representing the higher-order Markov chains (and HMMs), but the theory is not very satisfactory. The notation above was used by Paul Malcolm in his presentation to the IFORS Meeting in Melbourne in July 2011, "Dynamical recursive representation for order $M \geq 2$ discrete-time Markov chains, with applications to hidden-state estimation problems".

2.5 Exercises

Exercise 2.1 Consider a two-state chain with state space $\{e_1, e_2\}$. Suppose we observe a path $\{e_1, e_1, e_2, e_1, e_2, e_2, e_2, e_2, e_1, e_2, e_1, e_1, e_1, e_2, e_2\}$. What are estimates for $a_{11}, a_{12}, a_{21}, a_{22}$?

Exercise 2.2 Consider a two-state chain of order $M = 3$. Write down the form of the transition matrix for $X_{n+1} \otimes X_n \otimes X_{n-1}$.

3
Hidden Markov Models

3.1 Definitions

In this chapter a hidden Markov chain consists of two processes, a finite-state discrete-time Markov chain X as discussed in the first two chapters and a second finite-state discrete-time process Y whose transitions depend on the state of X. It is supposed that only Y is observed. From these observations of Y we wish to estimate the initial state of X, the transition probabilities of X and the probabilities describing how the transitions of Y depend on X. The first part of the chapter describes the basic estimation problems given the observations. The hidden Markov chain is then explicitly constructed on a canonical probability space by defining the appropriate likelihoods or densities. We then derive a recursion, or filter, for the hidden state given the observations. The chapter concludes by discussing the estimation of the likelihood function given the observations.

We begin by discussing the hidden Markov model as it is usually used in genomics, but expressed in our notation. Suppose there is the usual Markov chain $\{X_n\}$ with $X_n \in \{e_1, \dots, e_N\} \subset \mathbb{R}^N$. Its semi-martingale form is

$$X_{n+1} = A X_n + V_{n+1}$$

where $A = (a_{ji})$, and

$$a_{ji} = P(X_{n+1} = e_j \,|X_n = e_i).$$

Let $\pi_j = P(X_0 = e_j)$ denote the initial probabilities. However, we now suppose $\{X_n\}$ is not observed directly.

Rather we observe a second chain $\{Y_n\}$, where $Y_n \in \{f_1, f_2, \dots, f_M\} \subset$

$\mathbb{R}^M$ $\left(f_i = (0, 0, \ldots, 1, \ldots, 0)' \in \mathbb{R}^M\right)$. We call the probabilities

$$P(Y_n = f_k \mid X_n = e_i) = C_{ki}$$

emission probabilities. The quantity $C = (C_{ki})$ is the $M \times N$ matrix of emission probabilities.

Here we must have $C_{ki} \geq 0$ for all k, i and

$$\sum_{k=1}^{M} C_{ki} = 1.$$

This means that for a given choice $X_n = e_i$, the values Y_n are chosen from $\{f_1, \ldots, f_M\}$ according to fixed probabilities $\{C_{ki}, k = 1, 2, \ldots, M\}$. We shall comment on this below.

We can also write

$$Y_n = C\,X_n + W_n$$

where

$$\mathbf{E}[W_n \,|\mathcal{F}_n] = \mathbf{E}[W_n \,|X_n] = 0$$

for the sequence $\{Y_n\}$ of observations.

The two processes X and Y make up a hidden Markov model, HMM.

Four problems for Markov chains

There are four problems associated with hidden Markov models. These are discussed in some detail in the context of speech analysis in (Rabiner and Juang, 1993, Chapter 6) and in Jelinek (1997).

1. Scoring: Let $\lambda = (\pi, A, C)$ represent the parameters of the HMM. Find the probability

$$L_\theta = P(Y_0 = f_{j_0}, Y_1 = f_{j_1}, \ldots, Y_n = f_{j_n} \,|\lambda).$$

This is solved in Sectio 3.2.

2. Decoding: Given a set of observations

$$\{Y_0 = f_{i_0}, Y_1 = f_{i_1}, \ldots, Y_n = f_{i_n}\},$$

find $i_0^*, i_1^*, \ldots, i_n^*$ such that

$$P(X_0 = e_{i_0^*}, X_1 = e_{i_1^*}, \ldots, X_n = e_{i_n^*}, Y_0 = f_{j_0}, \ldots, Y_n = f_{j_n} \,|\lambda)$$

is a maximum. This gives the most likely path of the hidden Markov chain consistent with the observations. This is solved in Chapter 5.

3. Estimation: Based on observations of $Y_0, Y_1, \ldots, Y_n$, determine the parameters λ which maximize the score. This is solved in Chapter 6.

4. The best architecture for X: If X has no physical meaning, there may be many alternatives for $\{X_n\}$ and different choices for N. There are also issues of consistency. Suppose that we are given N, λ and we have simulated values for $\{Y_n\}$. We can now calibrate (that is estimate) N and λ given these simulated values. Consistency will mean that we return the original value of N (in particular) and the values of λ. These are discussed in Jelinek (1997).

Comments on the model

(a) Given $X_n = e_i$, the distribution of Y_n is iid. In other words,

$$P(Y_n = f_k) = C_{ki}$$

which does not depend on Y_{n-1} or other earlier terms in the sequence of observations.

(b) An alternative specification could be

$$P(Y_n = f_k \,|\, X_{n-1} = e_i) = C_{ki}$$

when we would use the semi-martingale decomposition[1]

$$Y_n = C\,X_{n-1} + W_n.$$

(c) We could propose an alternative model in which

$$Y_{n+1} = C(X_n)\,Y_n + W_{n+1}$$

or

$$Y_{n+1} = C(X_{n+1})\,Y_n + W_{n+1} \tag{3.1}$$

and various generalizations using a higher-order chain for the $\{Y_n\}$ modulated by the hidden chain $\{X_n\}$. Model (3.1) is discussed in Chapter 7.

We would need to consider the emission matrices of the form

$$C^i_{qp} = P(Y_{n+1} = f_q \,|Y_n = f_p, X_n = e_i).$$

[1] We could state here that it is a good practice to use an index on the term of a random process or sequence which corresponds to the time/location when its value is known.

for example. If these new emission probabilities have the same value for each p, then we obtain the previous model with

$$C_q^i = P(Y_{n+1} = f_q \,|X_n = e_i)\,.$$

(d) Jelinek (1997) considered models with emission probabilities of the form

$$C_{k,ji} = P(Y_n = f_k \mid X_n = e_j, X_{n-1} = e_i)$$

but showed on page 25 of his book, that these processes can be expressed in the original form, by enlarging the sate space of $\{X_n\}$ in the way we have already discussed using $Z_n = X_n \otimes X_{n-1}$ in place of (X_n, X_{n-1}).

We assume the following:

(i) $\{X_n\}$ is a Markov chain taking N values in $\{e_1, \ldots, e_N\} \subset \mathbb{R}^N$ but now not observed directly.

(ii) $\{Y_n\}$ is an observed chain taking values in $\{f_1, \ldots, f_M\} \subset \mathbb{R}^M$. Assume that there is a matrix C which is $M \times N$ so that

$$C_{ji} = P(Y_n = f_j \mid X_n = e_i).$$

The matrix C is called the emission matrix. Note $C_{ji} \geq 0$ and $\sum_{j=1}^M C_{ji} = 1$. The parameters of the model are now

$$\theta = (\pi(0), A, C)\,,$$

where

$$\pi_j(0) = P(X_0 = e_j), \quad A_{ji} = P(X_{n+1} = e_j \mid X_n = e_i),$$

and C is as above.

(iii) Given θ and given $X_0, X_1, \ldots, X_n$, the variables $Y_0, Y_1, \ldots, Y_n$ are independent.

Construction of the Hidden Markov Model

Remark We shall start $\{X_n\}, \{Y_n\}$ at $n = 0$. (Starting with $n = 1$ is popular in speech analysis. Also, $X_0, X_1, \ldots, Y_1, Y_2, \ldots$ is used.)

Example 3.1 In a simple situation related to genomics we could have $N = 2$ (coding, noncoding), $M = 4$.

Basic reference probability space Consider the reference probability space $(\Omega, \mathcal{F}, \overline{P})$, where

$$\Omega = \{\omega \,|\, \omega = (\omega_0, \omega_1, \ldots) \in \mathbb{R}^{N+M}\}$$

where for each i,

$$\omega_i = \begin{pmatrix} \omega_i^1 \\ \omega_i^2 \end{pmatrix}, \quad \omega_i^1 \in \{e_1, \ldots, e_N\}, \quad \omega_i^2 \in \{f_1, \ldots, f_M\}.$$

Consider subsets of Ω of the form

$$A = \{\omega \in \Omega \mid \omega_{i_1}^1 = e_{r_1}, \ldots, \omega_{i_n}^1 = e_{r_n}, \omega_{j_1}^2 = f_{s_1}, \ldots, \omega_{j_m}^2 = f_{s_m}\}.$$

For example, $A = \{\omega \in \Omega \,|\, \omega_0^1 = e_2, \omega_5^2 = f_3\}$. Again the σ-algebra $\mathcal{F}$ will be the smallest σ-agebra generated by all sets of this form.

For the set A just described, we set

$$\overline{P}(A) = \frac{1}{N^n} \cdot \frac{1}{M^m}.$$

Consider chains $\{X_n\}$ and $\{Y_n\}$ on this probability space obtained by setting

$$X_n(\omega) = \omega_n^1 \text{ and } Y_n(\omega) = \omega_n^2 \text{ for } n = 0, 1, 2, \ldots$$

Under this probability $\overline{P}$ the terms of the chain $\{X_n\}$ are iid with

$$\overline{P}(X_n = e_i) = \frac{1}{N} \text{ for each } n \text{ and } i.$$

and the terms of the chain $\{Y_n\}$ are iid with

$$\overline{P}(Y_n = f_i) = \frac{1}{M} \text{ for each } n \text{ and } i.$$

Furthermore, under $\overline{P}$ the two chains $\{X_n\}, \{Y_n\}$ are independent. This implies for example, that

$$\begin{aligned} \overline{P}(X_n = e_i, Y_m = f_j) &= \overline{P}(\{\omega \in \Omega \,|\, \omega_n^1 = e_i, \omega_m^2 = f_j\}) \\ &= \frac{1}{N} \cdot \frac{1}{M} = \overline{P}(X_n = e_i)\,\overline{P}(Y_m = f_j). \end{aligned}$$

Suppose we are given an $N \times N$ transition matrix $A = (A_{ji})$ with $A_{ji} \geq 0$ and $\sum_{j=1}^N A_{ji} = 1$ and an $M \times N$ emission matrix $C = (C_{ji})$ with $C_{ji} \geq 0$ and $\sum_{j=1}^M C_{ji} = 1$. Starting with $\overline{P}$ we shall now introduce a new probability P on $(\Omega, \mathcal{F})$ similarly to Chapter 1. However, under P in addition to X being a Markov chain with transition matrix A we wish Y to be related to X following the probabilities C_{ji}.

Definition 3.2 With $\{X_n\}$ and $\{Y_n\}$ defined as above, write

$$\overline{\lambda}_l = \begin{cases} NM\langle CX_l, Y_l\rangle\langle AX_{l-1}, X_l\rangle & \text{if } l \geq 1 \\ NM\langle CX_l, Y_l\rangle\langle \pi(0), X_l\rangle & \text{if } l = 0 \end{cases}.$$

Define the σ-algebras, or possible 'histories',

$$\mathcal{F}_n = \sigma\{X_0, \dots, X_n\}$$
$$\mathcal{Y}_n = \sigma\{Y_0, \dots, Y_n\}$$

and

$$\mathcal{G}_n = \sigma\{X_0, Y_0, X_1, Y_1, \dots, X_n, Y_n\} = \mathcal{F}_n \vee \mathcal{Y}_n.$$

Lemma 3.3 *For each* $l \geq 1$,

$$\overline{\mathbf{E}}\,[\overline{\lambda}_l \mid \mathcal{G}_{l-1}] = 1$$

and

$$\overline{\mathbf{E}}[\overline{\lambda}_0] = 1.$$

Proof Let $l \geq 1$.

$$\begin{aligned}
&\overline{\mathbf{E}}\,[NM\langle CX_l, Y_l\rangle\langle AX_{l-1}, X_l\rangle \mid \mathcal{G}_{l-1}] \\
&= \overline{\mathbf{E}}\left[\sum_{i=1}^{N}\sum_{j=1}^{M}\langle X_l, e_i\rangle\langle Y_l, f_j\rangle NM\langle CX_l, Y_l\rangle\langle AX_{l-1}, X_l\rangle \,\middle|\, \mathcal{G}_{l-1}\right] \\
&= NM\sum_{i=1}^{N}\sum_{j=1}^{M}\langle Ce_i, f_j\rangle\langle AX_{l-1}, e_i\rangle\,\overline{\mathbf{E}}[\langle X_l, e_i\rangle\langle Y_l, f_j\rangle \mid \mathcal{G}_{l-1}] \\
&= NM\sum_{i=1}^{N}\sum_{j=1}^{M}\langle Ce_i, f_j\rangle\langle AX_{l-1}, e_i\rangle\,\overline{\mathbf{E}}[\langle X_l, e_i\rangle\langle Y_l, f_j\rangle] \\
&= \sum_{i=1}^{N}\sum_{j=1}^{M}\langle Ce_i, f_j\rangle\langle AX_{l-1}, e_i\rangle \quad = \sum_{i=1}^{N}\langle Ce_i, 1\rangle\langle AX_{l-1}, e_i\rangle \\
&= \sum_{i=1}^{N}\langle AX_{l-1}, e_i\rangle \quad \text{as } \sum_{j=1}^{M} C_{ji} = 1 \\
&= \langle AX_{l-1}, 1\rangle \\
&= 1 \quad \text{as } \langle AX_{l-1}, 1\rangle = \mathbf{1}^\top AX_{l-1} = \mathbf{1}^\top X_{l-1} = 1.
\end{aligned}$$

The second identity is proved in a similar way. □

Definition 3.4 For $n \geq 0$, define

$$\overline{\Lambda}_n = \prod_{l=0}^{n} \overline{\lambda}_l$$

and P on $(\Omega, \mathcal{F})$ by

$$\left.\frac{dP}{d\overline{P}}\right|_{\mathcal{G}_n} = \overline{\Lambda}_n$$

for each $n = 0, 1, 2, \ldots$. We shall call $\overline{\Lambda}_n$ the likelihood function. The likelihood of the observations $y_{0:n}$ is the conditional expression of the likelihood function with respect to the observations $Y_{0:n} = y_{0:n}$ under the reference probability.

The construction of this new probability P uses Lemma 3.3 and follows the construction described in Chapter 1.

Lemma 3.5 *For $n \geq 0$,*

$$P(X_{n+1} = e_j \,|\, X_n = e_i) = A_{ji}.$$

when $P(X_n = e_i) \neq 0$.

Proof We first compute $P(X_{n+1} = e_j \ \& \ X_n = e_i)$. Then we obtain

$$P(X_n = e_i) = \sum_{j=1}^{N} P(X_{n+1} = e_j \ \& \ X_n = e_i),$$

from which we obtain the expression

$$P(X_{n+1} = e_j \,|\, X_n = e_i) = \frac{P(X_{n+1} = e_j \ \& \ X_n = e_i)}{P(X_n = e_i)}.$$

Now

$$\begin{aligned}
P(X_{n+1}e_j \ \& \ X_n = e_i) = {} & \mathbf{E}\left[\langle X_{n+1}, e_j\rangle\langle X_n, e_i\rangle\right] \\
= {} & \overline{\mathbf{E}}\left[\overline{\Lambda}_{n+1}\langle X_{n+1}, e_j\rangle\langle X_n, e_i\rangle\right] \\
= {} & NM\,\overline{\mathbf{E}}\left[\overline{\Lambda}_n\langle CX_{n+1}, Y_{n+1}\rangle\langle AX_n, X_{n+1}\rangle\langle X_{n+1}, e_j\rangle\langle X_n, e_i\rangle\right] \\
= {} & NM\,A_{ji}\,\overline{\mathbf{E}}\left[\overline{\Lambda}_n\langle Ce_j, Y_{n+1}\rangle\langle X_{n+1}, e_j\rangle\langle X_n, e_i\rangle\right] \\
= {} & NM\,A_{ji}\,\overline{\mathbf{E}}\left[\overline{\Lambda}_n\langle X_n, e_i\rangle\right]\overline{\mathbf{E}}\big[\langle Ce_i, Y_{n+1}\rangle\big]\overline{\mathbf{E}}\big[\langle X_{n+1}, e_j\rangle\big].
\end{aligned}$$

But

$$\overline{\mathbf{E}}\left[\langle Ce_i, Y_{n+1}\rangle\right] = \overline{\mathbf{E}}\left[\sum_{j=1}^{M}\langle Y_{n+1}, f_j\rangle C_{ji}\right] = \sum_{j=1}^{M} C_{ji}\overline{\mathbf{E}}\left[\langle Y_{n+1}, f_j\rangle\right]$$
$$= \sum_{j=1}^{M} C_{ji}\frac{1}{M} = \frac{1}{M}$$

and

$$\overline{\mathbf{E}}\left[\langle X_{n+1}, e_j\rangle\right] = \frac{1}{N}$$

so

$$P(X_{n+1} = e_j \;\&\; X_n = e_i) = A_{ji}\,\overline{\mathbf{E}}\left[\overline{\Lambda}_n\langle X_n, e_i\rangle\right],$$

and

$$P(X_n = e_i) = \overline{\mathbf{E}}\left[\overline{\Lambda}_n\langle X_n, e_i\rangle\right].$$

Thus

$$P(X_{n+1} = e_j \mid X_n = e_i) = A_{ji}.$$

holds in the case $P(X_n = e_i) \neq 0$. □

Lemma 3.6 *For $n \geq 0$,*

$$P\left(Y_n = f_j \mid X_n = e_i\right) = C_{ji}.$$

when $P(X_n = e_i) \neq 0$.

Proof For $n \geq 1$

$$\begin{aligned}
&P(Y_n = f_j \;\&\; X_n = e_i)\\
&\quad = \mathbf{E}\left[\langle Y_n, f_j\rangle\langle X_n, e_i\rangle\right]\\
&\quad = \overline{\mathbf{E}}\left[\overline{\Lambda}_n\langle Y_n, f_j\rangle\langle X_n, e_i\rangle\right]\\
&\quad = \overline{\mathbf{E}}\left[\overline{\Lambda}_{n-1}MN\langle CX_n, Y_n\rangle\langle AX_{n-1}, X_n\rangle\langle Y_n, f_j\rangle\langle X_n, e_i\rangle\right]\\
&\quad = C_{ji}MN\,\overline{\mathbf{E}}\left[\overline{\Lambda}_{n-1}\langle AX_{n-1}, e_i\rangle\langle Y_n, f_j\rangle\langle X_n, e_i\rangle\right]\\
&\quad = C_{ji}\,\overline{\mathbf{E}}\left[\overline{\Lambda}_{n-1}\langle AX_{n-1}, e_i\rangle\right]
\end{aligned}$$

as before. Thus (as the column sums of C are all 1)

$$P(X_n = e_i) = \sum_{j=1}^{M} P(Y_n = f_j \;\&\; X_n = e_i) = \overline{\mathbf{E}}\left[\overline{\Lambda}_{n-1}\langle AX_{n-1}, e_i\rangle\right],$$

and

$$P(Y_n = f_j \,|\, X_n = e_i) = \frac{P(Y_n = f_j \;\&\; X_n = e_i)}{P(X_n = e_i)} = C_{ji}. \qquad \square$$

Lemma 3.7 *For any $n \geq 1$,*

$$P(Y_0 = f_{i_0}, \ldots, Y_n = f_{i_n} \,|\, \mathcal{F}_n) = \prod_{j=0}^{n} P(Y_j = f_{i_j} \,|\, \mathcal{F}_n),$$

and further

$$P(Y_j = f_{i_j} \,|\, \mathcal{F}_n) = \langle CX_j, f_{i_j} \rangle.$$

This shows that, given the X values, the Y are iid.

Proof By the so-called Bayes' Identity (see Appendix C).

$$\mathbf{E}\left[\prod_{j=0}^{n} \langle Y_j, f_{i_j} \rangle \,\middle|\, \mathcal{F}_n \right] = \frac{\overline{\mathbf{E}}\left[\overline{\Lambda}_n \prod_{j=0}^{n} \langle Y_j, f_{i_j} \rangle \middle| \mathcal{F}_n \right]}{\overline{\mathbf{E}}\left[\overline{\Lambda}_n \,|\, \mathcal{F}_n \right]}.$$

We calculate the numerator

$$\begin{aligned}
\overline{\mathbf{E}}\left[\overline{\Lambda}_n \prod_{j=0}^{n} \langle Y_j, f_{i_j} \rangle \middle| \mathcal{F}_n \right] &= \overline{\mathbf{E}}\left[\prod_{l=0}^{n} \overline{\lambda}_l \prod_{j=0}^{n} \langle Y_j, f_{i_j} \rangle \,\middle|\, \mathcal{F}_n \right] \\
&= N^{n+1} M^{n+1} \left(\prod_{l=1}^{n} \langle AX_{l-1}, X_l \rangle \right) \langle \pi(0), X_0 \rangle \times \\
&\qquad \overline{\mathbf{E}}\left[\prod_{l=0}^{n} \langle CX_l, f_{i_l} \rangle \prod_{j=0}^{n} \langle Y_j, f_{i_j} \rangle \,\middle|\, \mathcal{F}_n \right] \\
&= N^{n+1} \prod_{l=1}^{n} \langle AX_{l-1}, X_l \rangle \langle \pi(0), X_0 \rangle \prod_{l=0}^{n} \langle CX_l, f_{i_l} \rangle.
\end{aligned}$$

We also have, since the column sums of C are 1,

$$\begin{aligned}
\overline{\mathbf{E}}\left[\overline{\Lambda}_n \,|\, \mathcal{F}_n \right] &= \sum_{i_0=1}^{M} \cdots \sum_{i_n=1}^{M} \overline{\mathbf{E}}\left[\overline{\Lambda}_n \prod_{j=0}^{n} \langle Y_j, f_{i_j} \rangle \,\middle|\, \mathcal{F}_n \right] \\
&= N^{n+1} \prod_{l=1}^{n} \langle AX_{l-1}, X_l \rangle \langle \pi(0), X_0 \rangle.
\end{aligned}$$

So

$$\mathbf{E}\left[\prod_{j=0}^{n} \langle Y_j, f_{i_j} \rangle \,\middle|\, \mathcal{F}_n \right] = \prod_{l=0}^{n} \langle CX_l, f_{i_l} \rangle.$$

However, for $0 \leq k \leq n$

$$\begin{aligned}
&P(Y_k = f_{i_k} \mid \mathcal{F}_n) \\
&= \sum_{i_1, i_2, \ldots, i_{k-1}, i_{k+1}, \ldots, i_n} P(Y_0 = f_{i_0}, \ldots, Y_{k-1} = f_{i_{k-1}}, \\
&\qquad\qquad Y_k = f_{i_k}, Y_{k+1} = f_{i_{k+1}}, \ldots, Y_n = f_{i_n} \mid \mathcal{F}_n) \\
&= \langle CX_k, f_{i_k} \rangle,
\end{aligned}$$

so

$$\mathbf{E}\left[\prod_{l=0}^{n} \langle Y_l, f_{i_l} \rangle \,\Big|\, \mathcal{F}_n\right] = \prod_{l=0}^{n} \mathbf{E}\left[\langle Y_l, f_{i_l} \rangle \mid \mathcal{F}_n\right]. \qquad \square$$

Now that we have provided a probabilistic framework for HMMs we can investigate various questions about them.

3.2 Calculation of the likelihood

Forward and Backward Algorithms

Given a set of observations the problem discussed in this chapter is how these can be used to estimate the likelihood. Both a forward and a backward algorithm are derived.

We define the likelihood of the observations $y_{0:n}$ by

$$L_\theta = P_\theta(Y_0 = y_0, \ldots, Y_n = y_n)$$

where $\theta = (\pi(0), A, 0)$ and $y_i \in \{f_1, f_2, \ldots, f_M\} \subset \mathbb{R}^M$. This expression is also given by

$$\sum_{i_0=1}^{N} \cdots \sum_{i_n=1}^{N} P_\theta(X_0 = e_{i_0}, Y_0 = y_0, \ldots, X_n = e_{i_n}, Y_n = y_n)$$

which has $N(n+1)$ terms. We shall present two algorithms for computing this expression:

(a) the forward algorithm;
(b) the backward algorithm.

Applications

(1) Given two sets of parameters θ_0 and θ_a we consider the following question: do the observations support a model with $\theta = \theta_0$ or with

$\theta = \theta_a$? We could calculate the likelihood under both alternatives. We would also need to develop criteria for rejecting one hypothesis in favor of an alternative.

(2) We could calculate the likelihood of the observation of a current model using parameters $\theta = \theta_0$ and compare this likelihood with the $\theta = \hat{\theta}$, the maximum likelihood parameter. This would be like testing $\theta = \theta_0$ versus $\theta \neq \theta_0$.

The Forward Algorithm

Write

$$\alpha_i(k) = P(Y_0 = y_0, \ldots, Y_k = y_k, X_k = e_i).$$

Then we have

$$P_\theta(Y_0 = y_0, \ldots, Y_k = y_k) = \sum_{i=1}^{N} \alpha_i(k).$$

We now obtain a recursion. We can write

$$\alpha_i(k) = \mathbf{E}_\theta \left[\langle Y_0, y_0 \rangle, \ldots, \langle Y_k, y_k \rangle \langle X_k, e_i \rangle \right].$$

Lemma 3.8 $\alpha(k+1) = B(y_{k+1}) \cdot A \cdot \alpha(k)$ *where* $B(y_{k+1})$ *is the* $N \times N$ *diagonal matrix*

$$\operatorname{diag}(\langle Ce_i, y_{k+1} \rangle) \quad \textit{and} \quad \alpha(0) = B(y_0)\pi(0).$$

Proof

$$\begin{aligned}
\alpha_i(k+1) &= \mathbf{E}_\theta[\langle Y_0, y_0\rangle, \dots, \langle Y_{k+1}, y_{k+1}\rangle\langle X_{k+1}, e_i\rangle] \\
&= \overline{\mathbf{E}}\left[\overline{\Lambda}_{k+1}\langle Y_0, y_0\rangle, \dots, \langle Y_{k+1}, y_{k+1}\rangle\langle X_{k+1}, e_i\rangle\right] \\
&= \overline{\mathbf{E}}\left[\overline{\Lambda}_k\overline{\lambda}_{k+1}\langle Y_0, y_0\rangle, \dots, \langle Y_{k+1}, y_{k+1}\rangle\langle X_{k+1}, e_i\rangle\right] \\
&= \overline{\mathbf{E}}\big[\overline{\Lambda}_k MN\langle AX_k, X_{k+1}\rangle\langle CX_{k+1}, Y_{k+1}\rangle\langle Y_0, y_0\rangle \\
&\qquad \cdots\langle Y_k, y_k\rangle\langle Y_{k+1}, y_{k+1}\rangle\langle X_{k+1}, e_i\rangle\big] \\
&= \overline{\mathbf{E}}\big[\overline{\Lambda}_k MN\langle AX_k, e_i\rangle\langle Ce_i, y_{k+1}\rangle\langle Y_0, y_0\rangle \\
&\qquad \cdots\langle Y_k, y_k\rangle\langle Y_{k+1}, y_{k+1}\rangle\langle X_{k+1}, e_i\rangle\big] \\
&= \overline{\mathbf{E}}\Bigg[\sum_{j=1}^{N}\langle X_k, e_j\rangle\overline{\Lambda}_k MN\langle AX_k, e_i\rangle\langle Ce_i, y_{k+1}\rangle\langle Y_0, y_0\rangle \\
&\qquad \cdots\langle Y_k, y_k\rangle\langle Y_{k+1}, y_{k+1}\rangle\langle X_{k+1}, e_i\rangle\Bigg] \\
&= \sum_{j=1}^{N} A_{ij}\,\langle Ce_i, y_{k+1}\rangle\,\overline{\mathbf{E}}\left[\overline{\Lambda}_k\langle Y_0, y_0\rangle, \dots, \langle Y_k, y_k\rangle\langle X_k, e_j\rangle\right] \\
&= \sum_{j=1}^{N} A_{ij}\,\langle Ce_i, y_{k+1}\rangle\alpha_j(k).
\end{aligned}$$

This forward algorithm is implemented by

$$\alpha(k+1) = B(y_{k+1})\cdot A\cdot\alpha(k),$$

where $B(y_{k+1})$ is the diagonal matrix $\operatorname{diag}\left(\langle Ce_i, y_{k+1}\rangle\right).$

We also need

$$\begin{aligned}
\alpha_i(0) &= P_\theta(Y_0 = y_0 \;\&\; X_0 = e_i) \\
&= \overline{\mathbf{E}}[MN\langle\pi(0), X_0\rangle\langle CX_0, Y_0\rangle\langle Y_0, y_0\rangle\langle X_0, e_i\rangle] \\
&= \pi(0)\langle Ce_i, y_0\rangle MN\,\overline{\mathbf{E}}\left[\langle Y_0, y_0\rangle\langle X_0, e_i\rangle\right] \\
&= \pi_i(0)\langle Ce_i, y_0\rangle;
\end{aligned}$$

that is,

$$\alpha(0) = B(y_0)\,\pi(0). \qquad \square$$

Therefore, the likelihood function is given by

$$\begin{aligned} L_\theta &= \sum_{i=1}^{N} \alpha_i(n) \\ &= \mathbf{1}^\top \alpha(n) \\ &= \mathbf{1}^\top B(y_n)\, A\, \alpha(n-1) \\ &= \mathbf{1}^\top B(y_n)\, A\, B(y_{n-1})\, A \cdots \alpha(0) \\ &= \mathbf{1}^\top B(y_n)\, A \cdots B(y_0)\, \pi(0) \end{aligned}$$

where $\mathbf{1}^\top = (1, 1, 1, \ldots, 1)$.

The Backward Algorithm

Let us now define

$$\beta_i(k) = P_\theta(Y_{k+1} = y_{k+1}, \ldots, Y_n = y_n \,|\, X_k = e_i)$$

and

$$\beta_i(n) = 1$$

for each $1 \leq i \leq N$.

Lemma 3.9 *For $k \leq n-1$,*

$$\beta(k) = A^\top B(y_{k+1})\, \beta(k+1) \text{ and } \beta(n) = \mathbf{1}$$

where $A^\top$ denotes the transpose of matrix A.

Proof For any $k \leq n-1$,

$$\begin{aligned} &\beta_i(k) \\ &= \mathbf{E}_\theta \left[\prod_{l=k+1}^{n} \langle Y_l, y_l \rangle \,\middle|\, X_k = e_i \right] \\ &= \sum_{j=1}^{N} \mathbf{E}_\theta \left[\prod_{l=k+1}^{n} \langle Y_l, y_l \rangle \,\middle|\, X_{k+1} = e_j, X_k = e_i \right] P\left(X_{k+1} = e_j \middle| X_k = e_i\right) \\ &= \sum_{j=1}^{N} A_{ji}\, \mathbf{E}_\theta \left[\prod_{l=k+1}^{n} \langle Y_l, y_l \rangle \,\middle|\, X_{k+1} = e_j \right]. \end{aligned} \tag{3.2}$$

To derive the second line of the above expression, we used

$$
\begin{aligned}
&P_\theta\left(Z = z \middle| X_k = e_i\right) \\
&= \frac{P_\theta\left(Z = z \,\&\, X_k = e_i\right)}{P_\theta(X_k = e_i)} \\
&= \frac{\sum_{j=1}^{N} P_\theta\left(Z = z \,\&\, X_{k+1} = e_j \,\&\, X_k = e_i\right)}{P_\theta(X_k = e_i)} \\
&= \frac{\sum_{j=1}^{N} P_\theta\left(Z = z \,\&\, X_{k+1} = e_j \,\&\, X_k = e_i\right)}{P_\theta(X_{k+1} = e_j \,\&\, X_k = e_i)} \frac{P_\theta(X_{k+1} = e_j \,\&\, X_k = e_i)}{P_\theta(X_k = e_i)} \\
&= \sum_{j=1}^{N} P_\theta\left(Z = z \,\middle|\, X_{k+1} = e_j \,\&\, X_k = e_i\right) P_\theta(X_{k+1} = e_j \,|\, X_k = e_i).
\end{aligned}
$$

We have shown in Lemma 3.7 that

$$
\mathbf{E}_\theta\left[\prod_{l=k+1}^{n} \langle Y_l, y_l\rangle \,\middle|\, \mathcal{F}_n\right] = \prod_{l=k+1}^{n} \mathbf{E}_\theta\left[\langle Y_l, y_l\rangle \,\middle|\, \mathcal{F}_n\right] = \prod_{l=k+1}^{n} \langle CX_l, y_l\rangle
$$

and so by the repeated conditioning

$$
\begin{aligned}
&\mathbf{E}_\theta\left[\prod_{l=k+1}^{n} \langle Y_l, y_l\rangle \,\middle|\, X_{k+1} = e_j\right] \\
&= \mathbf{E}_\theta\left[\mathbf{E}_\theta\left[\prod_{l=k+1}^{n} \langle Y_l, y_l\rangle \,\middle|\, X_{k+1} = e_j \,\&\, \mathcal{F}_n\right] \,\middle|\, X_{k+1} = e_j\right] \\
&= \mathbf{E}_\theta\left[\prod_{l=k+2}^{n} \langle CX_l, y_l\rangle \langle Ce_j, y_{k+1}\rangle \,\middle|\, X_{k+1} = e_j\right] \\
&= \langle Ce_j, y_{k+1}\rangle\, \mathbf{E}_\theta\left[\prod_{l=k+2}^{n} \langle CX_l, y_l\rangle \,\middle|\, X_{k+1} = e_j\right] \\
&= \langle Ce_j, y_{k+1}\rangle\, \mathbf{E}_\theta\left[\prod_{l=k+2}^{n} \langle Y_l, y_l\rangle \,\middle|\, X_{k+1} = e_j\right] \\
&= B_j(y_{k+1})\, \beta_j(k+1)\,.
\end{aligned}
$$

Substituting in (3.2) we obtain the recursion:

$$\begin{aligned}\beta_i(k) &= \sum_{j=1}^{N} A_{ji}\, B_j(y_{k+1})\, \beta_j(k+1) \\ &= \sum_{j=1}^{N} A_{ij}^{\top}\, B_j(y_{k+1})\, \beta_j(k+1) \\ &= \left(A^{\top} B(y_{k+1})\, \beta(k+1)\right)_i\end{aligned}$$

and so

$$\beta(k) = A^{\top}\, B(y_{k+1})\, \beta(k+1)\,.$$

To show that we can take $\beta(n) = \mathbf{1}$, we consider the calculations again for $k = n-1$. We use the arguments above, so we do not write all the steps. Then

$$\begin{aligned}\beta_i(n-1) &= \mathbf{E}_\theta\left[\langle Y_n, y_n\rangle \,\middle|\, X_{n-1} = e_i\right] \\ &= \sum_{j=1}^{n} A_{ji}\, \mathbf{E}_\theta\left[\langle Y_n, y_n\rangle \,\middle|\, X_n = e_j\right] = \sum_{j=1}^{n} A_{ji}\, \langle C\, e_j,\, y_j\rangle\end{aligned}$$

which says that

$$\beta(n-1) = A^{\top}\, B(y_n)\, \mathbf{1}$$

so we can take $\beta(n) = \mathbf{1}$, and we are done. □

Example 3.10 (After an example in Isaev (2006), page 38.) Let $M = 2$, $N = 3$

$$A = \begin{bmatrix} 0.3 & 0.2 & 0.4 \\ 0.3 & 0.4 & 0.3 \\ 0.4 & 0.4 & 0.3 \end{bmatrix},\quad C = \begin{bmatrix} 0.5 & 0.1 & 0.9 \\ 0.5 & 0.9 & 0.1 \end{bmatrix} \quad\text{and } \pi(0) = \begin{bmatrix} 0.2 \\ 0.3 \\ 0.5 \end{bmatrix}.$$

Suppose we observe f_1, f_1, f_2.

We first compute the likelihood using the forward algorithm. We shall present answers to four decimal places. We know that

$$\begin{aligned}\alpha(0) &= B(y_0)\, \pi(0) \\ &= \begin{bmatrix} 0.5 & 0 & 0 \\ 0 & 0.1 & 0 \\ 0 & 0 & 0.9 \end{bmatrix} \cdot \begin{bmatrix} 0.2 \\ 0.3 \\ 0.5 \end{bmatrix} = \begin{bmatrix} 0.10 \\ 0.03 \\ 0.45 \end{bmatrix}\end{aligned}$$

and using $\alpha(0)$,

$$\begin{aligned}\alpha(1) &= B(y_1)A\,\alpha(0)\\ &= \begin{bmatrix} 0.5 & 0 & 0 \\ 0 & 0.1 & 0 \\ 0 & 0 & 0.9 \end{bmatrix} \cdot A \cdot \begin{bmatrix} 0.10 \\ 0.03 \\ 0.45 \end{bmatrix} = \begin{bmatrix} 0.1080 \\ 0.0177 \\ 0.1683 \end{bmatrix}\end{aligned}$$

which then leads to

$$\begin{aligned}\alpha(2) &= B(y_2)A\,\alpha(1)\\ &= \begin{bmatrix} 0.5 & 0 & 0 \\ 0 & 0.9 & 0 \\ 0 & 0 & 0.1 \end{bmatrix} \cdot A \cdot \begin{bmatrix} 0.1080 \\ 0.0177 \\ 0.1683 \end{bmatrix} = \begin{bmatrix} 0.0516 \\ 0.0810 \\ 0.0101 \end{bmatrix}.\end{aligned}$$

Thus, the answer is given by

$$P_\theta(f_1 f_1 f_2) = \mathbf{1}^\top \alpha(2) = 0.1427\,.$$

We used MATLAB and $B(y_1)$ was entered as `diag`([0.5 0.9 0.1]).

Now we compute the likelihood using backward algorithm. We let

$$\beta(2) = \begin{bmatrix} 1 \\ 1 \\ 1 \end{bmatrix}.$$

Then

$$\beta(1) = A^\top \cdot \begin{bmatrix} 0.5 & 0 & 0 \\ 0 & 0.9 & 0 \\ 0 & 0 & 0.1 \end{bmatrix} \cdot \beta(2) = \begin{bmatrix} 0.46 \\ 0.50 \\ 0.50 \end{bmatrix}$$

and so

$$\beta(0) = A^\top B(y_1)\beta(1) = \begin{bmatrix} 0.264 \\ 0.246 \\ 0.242 \end{bmatrix}.$$

Therefore, we have with $n = 2$ and $N = 3$,

$$
\begin{aligned}
P_\theta(f_1 f_1 f_2) &= P_\theta(Y_0 = y_0, \dots, Y_n = y_n) \\
&= \sum_{i=1}^{N} P_\theta(Y_0 = y_0, \dots, Y_n = y_n \,\&\, X_0 = e_i) \\
&= \sum_{i=1}^{N} P_\theta\left(Y_0 = y_0, \dots \,\&\, Y_n = y_n \right| X_0 = e_i)\, \pi_i(0) \\
&= \sum_{i=1}^{N} P_\theta(Y_0 = y_0 | \, X_0 = e_i)\, P_\theta(Y_1 = y_1, \dots, Y_n = y_n | \, X_0 = e_i)\, \pi_i(0) \\
&= \sum_{i=1}^{N} B_i(y_0)\, \beta_i(0)\, \pi_i(0) \\
&= \pi^\top(0)\, B(y_0)\, \beta(0) \\
&= \begin{bmatrix} 0.2 & 0.3 & 0.5 \end{bmatrix} \begin{bmatrix} 0.5 & 0 & 0 \\ 0 & 0.1 & 0 \\ 0 & 0 & 0.9 \end{bmatrix} \begin{bmatrix} 0.264 \\ 0.246 \\ 0.242 \end{bmatrix} = 0.1427.
\end{aligned}
$$

This is the same answer as obtained by using the forward algorithm.

Using the backward recursion we can express the likelihood function

$$
\begin{aligned}
L_\theta &= \pi^\top(0)\, B(y_0)\, \beta(0) \\
&= \pi^\top(0)\, B(y_0)\, A^\top B(y_1)\, \beta(1) \\
&= \pi^\top(0)\, B(y_0)\, A^\top B(y_1) \cdots A^\top B(y_n) 1.
\end{aligned}
$$

Using the forward recursion

$$
\begin{aligned}
L_\theta &= \mathbf{1}^\top \alpha(n) \\
&= \mathbf{1}^\top B(y_n)\, A\, \alpha(n-1) \\
&= \mathbf{1}^\top B(y_n)\, A \cdots B(y_1)\, A\, \alpha(0) \\
&= \mathbf{1}^\top B(y_n)\, A \cdots B(y_1)\, A\, B(y_0)\, \pi(0)
\end{aligned}
$$

which is just the transpose of the expression above. Because L_θ is scalar, this means that both calculations give the same answer.

We can summarize the above results as follows.

Lemma 3.11 *We have*

$$
\begin{aligned}
&\alpha(0) = B(y_0)\pi(0) \\
&\alpha(k) = B(y_k)\, A\, B(y_{k-1})\, A \cdots B(y_0)\, \pi(0) \quad \textit{for } k \ge 1
\end{aligned}
$$

and

$$\begin{aligned}\beta(n) &= \mathbf{1}\\ \beta(k) &= A^\top B(y_{k+1})\, A^\top\, B(y_{k+2}) \cdots A^\top\, B(y_n)\, \mathbf{1} \quad \textit{for } 0 \le k \le n-1.\end{aligned}$$

Because

$$\beta(k)^\top = \mathbf{1}^\top B(y_n) A\, B(y_{n-1}) A \cdots B(y_{k+1}) A$$

we have for any $0 \le k \le n$,

$$L_\theta = \beta(k)^\top\, \alpha(k) = \sum_{i=1}^{N} \beta_i(k)\, \alpha_i(k)\,.$$

Remark We note that the forward algorithm is usually used to compute the likelihood, but in what follows there are useful applications from knowing the solution of the backward recursion as well.

3.3 Exercises

Exercise 3.1 State the analogous results to Lemmas 3.3, 3.5 and 3.6 so that the dynamics of the chain and observation process are

$$\begin{aligned}X_{n+1} &= AX_n + V_{n+1}\\ Y_n &= CX_{n-1} + W_n.\end{aligned}$$

Exercise 3.2 Rework Example 3.10 with

$$A = \begin{pmatrix} 0.5 & 0.3 & 0.2\\ 0.3 & 0.3 & 0.5\\ 0.2 & 0.4 & 0.3\end{pmatrix}$$

$$C = \begin{pmatrix} 0.4 & 0.8 & 0.2\\ 0.6 & 0.2 & 0.8\end{pmatrix} \qquad \pi(0) = \begin{pmatrix} 0.3\\ 0.3\\ 0.4\end{pmatrix}$$

and an observed sequence

$$f_2, f_2, f_1, f_2\,.$$

4
Filters and Smoothers

4.1 Introduction

A filter is a recursive estimate of the hidden state X given observations of Y up to the present time. A smoother is an estimate of the hidden state at a given time given observations of Y to some future time. This chapter obtains expressions for these two estimates.

We shall compute the filters

$$\hat{X}_{k|k} = \mathbf{E}[X_k | \mathcal{Y}_k]$$

from which

$$P(X_k = e_i \,\big|\, \mathcal{Y}_k) = \langle \hat{X}_{k|k} \, , e_i \rangle \, .$$

We shall also compute smoothers

$$\hat{X}_{k|n} = \mathbf{E}[X_k | \mathcal{Y}_n]$$

for $0 \leq k \leq n$. The expression for the filter can be obtained from the smoother when $k = n$.

From now on we shall use the abbreviation $Y_{0:n}$ for the sequence $(Y_0, Y_1, \ldots, Y_n)$ and $Y_{0:n} = y_{0:k}$ will stand for $Y_0 = y_0, Y_1 = y_1, \ldots, Y_k = y_k$ where $y_{0:k}$ is a sequence of specific values $(y_0, y_1, \ldots, y_k)$ with each y_l equal to some $f_i = (0, \ldots, 0, 1, 0, \ldots, 0)^\top$.

Lemma 4.1 *For $0 \leq k \leq n$ and $1 \leq i \leq N$,*

$$P(X_k = e_j \,|\, Y_{0:n} = y_{0:n}) = \frac{\alpha_j(k)\, \beta_j(k)}{\sum_{i=1}^N \alpha_i(k)\, \beta_i(k)} \equiv \frac{\alpha_j(k)\, \beta_j(k)}{L_\theta} \, .$$

Note that

$$L_\theta = \sum_{i=1}^N \alpha_i(k) \beta_i(k)$$

does not depend on k.

In order to prove this result we need the next lemma which holds for a general Markov chain.

Lemma 4.2 *Let $X = \{X_m : m = 0, 1, 2, \ldots\}$ be a Markov chain. Let $0 \le k \le n$ and let e be an element of the state space of X. Then given the event $X_k = e$,*

$$(X_0, X_1, \ldots, X_{k-1}) \quad \text{and} \quad (X_{k+1}, X_{k+2}, \ldots, X_n)$$

are independent.

Proof Let $x_0, x_1, \ldots, x_{k-1}, x_{k+1}, \ldots, x_n$ be elements of the state space of X which we take as the usual unit vectors in some $\mathbb{R}^N$, say. Then

$$\begin{aligned}
&P(X_0 = x_0, \ldots, X_{k-1} = x_{k-1}, X_{k+1} = x_{k+1}, \ldots, X_n = x_n \,|\, X_k = e) \\
&\quad = \mathbf{E}\left[\prod_{l=0}^{k-1} \langle X_l, x_l \rangle \prod_{l=k+1}^{n} \langle X_l, x_l \rangle \,\middle|\, X_k = e \right] \\
&\quad = \mathbf{E}\left[\mathbf{E}\left[\prod_{l=0}^{k-1} \langle X_l, x_l \rangle \prod_{l=k+1}^{n} \langle X_l, x_l \rangle \,\middle|\, \mathcal{F}_{k-1} \,\&\, X_k = e \right] \,\middle|\, X_k = e \right] \\
&\quad = \mathbf{E}\left[\prod_{l=0}^{k-1} \langle X_l, x_l \rangle \, \mathbf{E}\left[\prod_{l=k+1}^{n} \langle X_l, x_l \rangle \,\middle|\, \mathcal{F}_{k-1} \,\&\, X_k = e \right] \,\middle|\, X_k = e \right] \\
&\quad = \mathbf{E}\left[\prod_{l=0}^{k-1} \langle X_l, x_l \rangle \, \mathbf{E}\left[\prod_{l=k+1}^{n} \langle X_l, x_l \rangle \,\middle|\, X_k = e \right] \,\middle|\, X_k = e \right] \\
&\quad = \mathbf{E}\left[\prod_{l=0}^{k-1} \langle X_l, x_l \rangle \,\middle|\, X_k = e \right] \mathbf{E}\left[\prod_{l=k+1}^{n} \langle X_l, x_l \rangle \,\middle|\, X_k = e \right] \\
&\quad = P(X_0 = x_0, \ldots, X_{k-1} = x_{k-1} \,|\, X_k = e) \times \\
&\qquad\qquad P(X_{k+1} = x_{k+1}, \ldots, X_n = x_n \,|\, X_k = e)
\end{aligned}$$

and the lemma is proved. Going from line four to line five we used the Markov property of the chain. □

Proof of Lemma 4.1 For any $0 \le k \le n$,

$$\begin{aligned}
&P_\theta(X_k = e_j | \, Y_0 = y_0 \,, \ldots, Y_n = y_n) \\
&\quad = \frac{P_\theta(Y_0 = y_0, \ldots, Y_n = y_n \,\&\, X_k = e_j)}{P_\theta(Y_0 = y_0 \,, \ldots, Y_n = y_n)} \\
&\quad = \frac{P_\theta(Y_0 = y_0, \ldots, Y_n = y_n \,\&\, X_k = e_j)}{L_\theta} .
\end{aligned}$$

The numerator here is

$$
\begin{aligned}
&\mathbf{E}_\theta\left[\prod_{l=0}^{n}\langle Y_l, y_l\rangle\langle X_k, e_j\rangle\right]\\
&= \mathbf{E}_\theta\left[\mathbf{E}\left[\prod_{l=0}^{n}\langle Y_l, y_l\rangle\langle X_k, e_j\rangle \,\middle|\, \mathcal{F}_n\right]\right]\\
&= \mathbf{E}_\theta\left[\langle X_k, e_j\rangle\,\mathbf{E}\left[\prod_{l=0}^{n}\langle Y_l, y_l\rangle \,\middle|\, \mathcal{F}_n\right]\right]\\
&= \mathbf{E}_\theta\left[\langle X_k, e_j\rangle \prod_{l=0}^{n}\langle CX_l, y_l\rangle\right]\\
&= P_\theta(X_k = e_j)\,\langle Ce_j, y_k\rangle\,\mathbf{E}_\theta\left[\prod_{l=0, l\neq k}^{n}\langle CX_l, y_l\rangle \,\middle|\, X_k = e_j\right]
\end{aligned}
$$

where we used Lemma 3.7 to go from line 2 to line 3. By Lemma 4.2 we conclude that

$$
\begin{aligned}
&\mathbf{E}_\theta\left[\prod_{l=0, l\neq k}^{n}\langle CX_l, y_l\rangle \,\middle|\, X_k = e_j\right]\\
&\quad= \mathbf{E}_\theta\left[\prod_{l=0}^{k-1}\langle CX_l, y_l\rangle \,\middle|\, X_k = e_j\right]\mathbf{E}_\theta\left[\prod_{l=k+1}^{n}\langle CX_l, y_l\rangle \,\middle|\, X_k = e_j\right]
\end{aligned}
$$

and hence

$$
\begin{aligned}
&\mathbf{E}_\theta\left[\prod_{l=0}^{n}\langle Y_l, y_l\rangle\langle X_k, e_j\rangle\right]\\
&= P_\theta(X_k = e_j)\,\mathbf{E}_\theta\left[\prod_{l=0}^{k}\langle CX_l, y_l\rangle \,\middle|\, X_k = e_j\right] \times\\
&\qquad \mathbf{E}_\theta\left[\prod_{l=k+1}^{n}\langle CX_l, y_l\rangle \,\middle|\, X_k = e_j\right]\\
&= P_\theta(X_k = e_j)\,\mathbf{E}_\theta\left[\prod_{l=0}^{k}\langle Y_l, y_l\rangle \,\middle|\, X_k = e_j\right]\mathbf{E}_\theta\left[\prod_{l=k+1}^{n}\langle Y_l, y_l\rangle \,\middle|\, X_k = e_j\right]\\
&= \mathbf{E}_\theta\left[\prod_{l=0}^{k}\langle Y_l, y_l\rangle\,\langle X_k, e_j\rangle\right]\mathbf{E}_\theta\left[\prod_{l=k+1}^{n}\langle Y_l, y_l\rangle \,\middle|\, X_k = e_j\right] = \alpha_j(k)\,\beta_j(k)
\end{aligned}
$$

and this with Lemma 3.11 proves the result. □

As a corollary we have the filter result.

Lemma 4.3 *We have for any $k \geq 0$,*

$$P(X_k = e_j | Y_{0:k} = y_{0:k}) = \frac{\alpha_j(k)}{\sum_{j=1}^{N} \alpha_j(k)}.$$

Proof We obtain this result by setting $k = n$ in Lemma 4.1 and then note that n could be taken to have any integer value $n \geq 0$. From the definition of $\alpha(k)$, it depends only on $y_{0:k}$. □

4.2 Decoding

Given the observations $Y_{0:n} = y_{0:n}$ we seek the sequence

$$i_0^*, i_1^*, i_2^*, \ldots, i_n^*$$

so that (given θ)

$$P_\theta \left(X_0 = e_{i_0^*}, Y_0 = y_0, X_1 = e_{i_1^*}, Y_1 = y_1, \ldots, X_n = e_{i_n^*}, Y_n = y_n\right)$$

is maximal. The solving of this problem is often called decoding.

One approach to decoding is to use the expression for the smoothers

$$P(X_k = e_j | Y_{0:n} = y_{0:n}) = \frac{\alpha_j(k)\, \beta_j(k)}{\sum_{i=1}^{N} \alpha_i(k)\, \beta_i(k)} \tag{4.1}$$

and for each k with $0 \leq k \leq n$ find the value of j which maximizes the expression in (4.1) and call the answer i_k^*. This is a reasonably straightforward approach to decoding, but the problem with it is that

$$e_{i_0^*}, e_{i_1^*}, e_{i_2^*}, \ldots, e_{i_n^*}$$

may not be attainable. This is discussed in (Rabiner and Juang, 1993, Chapter 6). It could happen in the estimation of A that for some j, i that $A_{ji} = 0$. However if $A_{ji} > 0$ for all j, i, then this method can be used. In general the Viterbi algorithm is used to solve this decoding problem. This will be discussed below.

4.3 Further remarks on filters and smoothers

We follow the approach in Elliott et al. (1995). From the Bayes conditional expectation formula (see Appendix 3)

$$\hat{X}_{k|k} = \mathbf{E}_\theta\left[X_k \,|\, \mathcal{Y}_k\right] = \frac{\overline{\mathbf{E}}\left[\overline{\Lambda}_k \, X_k \,|\, \mathcal{Y}_k\right]}{\overline{\mathbf{E}}\left[\overline{\Lambda}_k \,|\, \mathcal{Y}_k\right]}\,.$$

Defining

$$q_k = \overline{\mathbf{E}}\left[\overline{\Lambda}_k \, X_k \,|\, \mathcal{Y}_k\right]$$

we have

$$\hat{X}_{k|k} = \frac{q_k}{\langle \mathbf{1}, q_k \rangle}\,.$$

We now provide a recursion formula for $\{q_k\}$.

Theorem 4.4

$$q_{k+1} = M B(Y_{k+1}) A q_k\,.$$

Proof Firstly,

$$\begin{aligned}
q_0 &= \overline{\mathbf{E}}\left[\overline{\Lambda}_0 \, X_0 \,|\, Y_0\right] = M\,N\,\overline{\mathbf{E}}\left[\langle C\,X_0, Y_0\rangle \,\langle \pi(0), X_0\rangle \, X_0 \,\big|\, Y_0\right] \\
&= \sum_{i=1}^{N} M\,N\,\overline{\mathbf{E}}\left[\langle X_0, e_i\rangle \,\langle C\,X_0, Y_0\rangle \,\langle \pi(0), X_0\rangle \, X_0 \,\big|\, Y_0\right] \\
&= \sum_{i=1}^{N} M\,N\,\langle C\,e_i, Y_0\rangle \,\langle \pi(0), e_i\rangle \, e_i \,\overline{\mathbf{E}}\left[\langle X_0, e_i\rangle\right] \\
&= \sum_{i=1}^{N} M\,\langle C\,e_i, Y_0\rangle \,\pi_i(0)\, e_i \\
&= M\,B(Y_0)\,\pi(0).
\end{aligned}$$

We note that q_0 is a random variable, because it is a function of the random variable Y_0. However when $Y_0 = y_0$ we obtain a real number which is the same as $\alpha(0)$.

For a recursion, suppose $k \geq 0$. Then

$$\begin{aligned}
&q_{k+1} \\
&= \overline{\mathbf{E}}\left[\overline{\Lambda}_{k+1} \, X_{k+1} \,|\, \mathcal{Y}_{k+1}\right] \\
&= \overline{\mathbf{E}}\left[\overline{\Lambda}_k \, \overline{\lambda}_k \, X_{k+1} \,|\, \mathcal{Y}_{k+1}\right] \\
&= M\,N\,\overline{\mathbf{E}}\left[\overline{\Lambda}_k \,\langle C\,X_{k+1}, Y_{k+1}\rangle \,\langle A\,X_k, X_{k+1}\rangle \, X_{k+1} \,|\, \mathcal{Y}_{k+1}\right]
\end{aligned}$$

$$= M\,N \sum_{j=1}^{N} \overline{\mathbf{E}}\left[\langle X_{k+1}, e_j\rangle\, \overline{\Lambda}_k \,\langle C\,X_{k+1}, Y_{k+1}\rangle\,\langle A\,X_k, X_{k+1}\rangle\, X_{k+1} \big|\, \mathcal{Y}_{k+1}\right]$$

$$= M\,N \sum_{j=1}^{N} \overline{\mathbf{E}}\left[\langle X_{k+1}, e_j\rangle\, \overline{\Lambda}_k \,\langle C\,e_j, Y_{k+1}\rangle\,\langle A\,X_k, e_j\rangle\, e_j \big|\, \mathcal{Y}_{k+1}\right]$$

$$= M\,N \sum_{j=1}^{N} \langle C\,e_j, Y_{k+1}\rangle\, \overline{\mathbf{E}}\left[\langle X_{k+1}, e_j\rangle\, \overline{\Lambda}_k \,\langle A\,X_k, e_j\rangle\, e_j \big|\, \mathcal{Y}_{k+1}\right]$$

$$= M\,N \sum_{j=1}^{N} \langle C\,e_j, Y_{k+1}\rangle\, \overline{\mathbf{E}}\left[\langle X_{k+1}, e_j\rangle\, \overline{\Lambda}_k \,\langle A\,X_k, e_j\rangle\, e_j \big|\, \mathcal{Y}_{k}\right]$$

$$= M \sum_{j=1}^{N} \langle C\,e_j, Y_{k+1}\rangle\, \overline{\mathbf{E}}\left[\overline{\Lambda}_k \,\langle A\,X_k, e_j\rangle\, e_j \big|\, \mathcal{Y}_{k}\right]$$

$$= M \sum_{j=1}^{N} \langle C\,e_j, Y_{k+1}\rangle\, \left\langle A\,\overline{\mathbf{E}}\left[\overline{\Lambda}_k \,X_k \big|\, \mathcal{Y}_{k}\right], e_j\right\rangle\, e_j$$

$$= M \sum_{j=1}^{N} \langle C\,e_j, Y_{k+1}\rangle\, \langle A\,q_k, e_j\rangle\, e_j\,.$$

This implies that

$$q_{k+1} = M\,B\,(Y_{k+1})\;A\,q_k\,. \qquad \square$$

Of course $\{q_k\}$ is a sequence of random variables. For a set of observations $y_{0:k}$ we obtain the sequence $\{\alpha(k)\}$, the recursion for which was

$$\alpha(k+1) = B\,(y_{k+1})\;A\,\alpha(k).$$

We now consider the smoothers. We shall use the notation

$$\overline{\Lambda}_{r,s} = \prod_{l=r}^{s} \overline{\lambda}_l$$

for $r \le s$.

For $0 \le k \le n$,

$$\hat{X}_{k|n} = \mathbf{E}_\theta\left[X_k \big|\, \mathcal{Y}_n\right] = \frac{\overline{\mathbf{E}}\left[\overline{\Lambda}_n\,X_k \big|\, \mathcal{Y}_n\right]}{\overline{\mathbf{E}}\left[\overline{\Lambda}_n \big|\, \mathcal{Y}_n\right]} = \frac{\overline{\mathbf{E}}\left[\overline{\Lambda}_n\,X_k \big|\, \mathcal{Y}_n\right]}{\left\langle \mathbf{1}, \overline{\mathbf{E}}\left[\overline{\Lambda}_n\,X_k \big|\, \mathcal{Y}_n\right]\right\rangle}$$

and

$$\overline{\mathbf{E}}\left[\overline{\Lambda}_n X_k \middle| \mathcal{Y}_n\right] = \overline{\mathbf{E}}\left[\overline{\mathbf{E}}\left[\overline{\Lambda}_n X_k \,\middle|\, \mathcal{Y}_n \vee \mathcal{F}_k\right] \,\middle|\, \mathcal{Y}_n\right]$$
$$= \overline{\mathbf{E}}\left[\overline{\Lambda}_k X_k \overline{\mathbf{E}}\left[\overline{\Lambda}_{k+1,n} \,\middle|\, \mathcal{Y}_n \vee \mathcal{F}_k\right] \,\middle|\, \mathcal{Y}_n\right].$$

By the Markov property,

$$\overline{\mathbf{E}}\left[\overline{\Lambda}_{k+1,n} \,\middle|\, \mathcal{Y}_n \vee \mathcal{F}_k\right] = \overline{\mathbf{E}}\left[\overline{\Lambda}_{k+1,n} \,\middle|\, \mathcal{Y}_n \vee X_k\right].$$

Write

$$v_{k,n} = \left(v_{k,n}^1, v_{k,n}^2, \dots, v_{k,n}^N\right)^\top$$

where

$$v_{k,n}^i = \overline{\mathbf{E}}\left[\overline{\Lambda}_{k+1,n} \,\middle|\, \mathcal{Y}_n \vee \{X_k = e_i\}\right].$$

We shall derive a backward recursion for $\{v_{k,n} \,|\, 0 \leq k \leq n\}$.

Theorem 4.5 *For $k = n$, $v_{n,n} = \mathbf{1}$ and for $0 \leq k \leq n-1$,*

$$v_{k,n} = M\, A^\top B\left(Y_{k+1}\right) v_{k+1,n}.$$

Remark The expression for

$$\mathbf{E}\left[Z \middle| Y_{0:k} = y_{0:k}\right]$$

for any random variable Z is obtained as follows. By the so-called Doob–Dynkin Lemma (seeOksendal, 2010, Lemma 2.1.2),

$$\mathbf{E}\left[Z \middle| Y_{0:k}\right] = h(Y_0, \dots, Y_k)$$

for some Borel function

$$h : \mathbb{R}^k \to \mathbb{R}.$$

Then

$$\mathbf{E}\left[Z \middle| Y_{0:k} = y_{0:k}\right] = h(y_0, \dots, y_k).$$

Remark This sequence $\{v_{k,n}\}$ is a sequence of random variables and when $Y_{0:n} = y_{0:n}$ then $v_{k,n}$ is the same as $\beta(k)$ considered earlier.

Proof For $1 \leq i \leq N$ and $0 \leq k \leq n-2$,

$$\begin{aligned}
&v_{k,n}^i \\
&= \overline{\mathbf{E}}\left[\overline{\Lambda}_{k+1,n} \,\middle|\, \mathcal{Y}_n \vee \{X_k = e_i\}\right] \\
&= \overline{\mathbf{E}}\left[\overline{\Lambda}_{k+2,n}\, \overline{\lambda}_{k+1} \,\middle|\, \mathcal{Y}_n \vee \{X_k = e_i\}\right] \\
&= M\, N\, \overline{\mathbf{E}}\left[\overline{\Lambda}_{k+2,n} \left\langle C\, X_{k+1}, Y_{k+1}\right\rangle \left\langle A\, X_k, X_{k+1}\right\rangle \,\middle|\, \mathcal{Y}_n \vee \{X_k = e_i\}\right]
\end{aligned}$$

$$= M\,N \sum_{j=1}^{N} \overline{\mathbf{E}}\big[\langle X_{k+1}, e_j\rangle\, \overline{\Lambda}_{k+2,n}\, \langle C\, X_{k+1}, Y_{k+1}\rangle$$

$$\langle A\, X_k, X_{k+1}\rangle \,\big|\, \mathcal{Y}_n \vee \{X_k = e_i\}\big]$$

$$= M\,N \sum_{j=1}^{N} \langle C\, e_j, Y_{k+1}\rangle\, A_{ji}\, \overline{\mathbf{E}}\left[\langle X_{k+1}, e_j\rangle\, \overline{\Lambda}_{k+2,n} \,\middle|\, \mathcal{Y}_n \vee \{X_k = e_i\}\right]$$

$$= M \sum_{j=1}^{N} \langle C\, e_j, Y_{k+1}\rangle\, A_{ji}\, \overline{\mathbf{E}}\left[\overline{\Lambda}_{k+2,n} \,\middle|\, \mathcal{Y}_n \vee \{X_k = e_i\} \vee \{X_{k+1} = e_j\}\right]$$

$$= M \sum_{j=1}^{N} \langle C\, e_j, Y_{k+1}\rangle\, A_{ji}\, \overline{\mathbf{E}}\left[\overline{\Lambda}_{k+2,n} \,\middle|\, \mathcal{Y}_n \vee \{X_{k+1} = e_j\}\right]$$

$$= M \sum_{j=1}^{N} \langle C\, e_j, Y_{k+1}\rangle\, A_{ji}\, v^j_{k+1,n} = M \sum_{j=1}^{N} A^\top_{ij} B_{jj}(Y_{k+1})\, v^j_{k+1,n}\,.$$

Therefore, the recursion holds for $0 \le k \le n-2$. When $k = n-1$ we have

$$v^i_{n-1,n}$$

$$= \overline{\mathbf{E}}\left[\Lambda_{n,n} \,\middle|\, \mathcal{Y}_n \vee \{X_{n-1} = e_i\}\right]$$

$$= M\,N\, \overline{\mathbf{E}}\left[\langle C\, X_n, Y_n\rangle\, \langle A\, X_{n-1}, X_n\rangle \,\middle|\, \mathcal{Y}_n \vee \{X_{n-1} = e_i\}\right]$$

$$= M\,N \sum_{j=1}^{N} \overline{\mathbf{E}}\left[\langle X_n, e_j\rangle\, \langle C\, X_n, Y_n\rangle\, \langle A\, X_{n-1}, X_n\rangle \,\middle|\, \mathcal{Y}_n \vee \{X_{n-1} = e_i\}\right]$$

$$= M\,N \sum_{j=1}^{N} \overline{\mathbf{E}}\left[\langle X_n, e_j\rangle\, \langle C\, X_n, Y_n\rangle\, \langle A\, X_{n-1}, X_n\rangle \,\middle|\, \mathcal{Y}_n \vee \{X_{n-1} = e_i\}\right]$$

$$= M\,N \sum_{j=1}^{N} \langle C\, e_j, Y_n\rangle\, A_{ji}\, \overline{\mathbf{E}}\left[\langle X_n, e_j\rangle \,\middle|\, \mathcal{Y}_n \vee \{X_{n-1} = e_i\}\right]$$

$$= M \sum_{j=1}^{N} \langle C\, e_j, Y_n\rangle\, A_{ji}\,.$$

This is the same as $v_{n-1,n} = M\, A^\top B(Y_n)\, \mathbf{1}$ implying we should take $v_{n,n} = \mathbf{1}$. The lemma is proved. □

We can write down explicit expressions for $v_{k,n}$ and q_k. In fact

$$q_k = M^{k+1}\, B(Y_k)\, A\, B(Y_{k-1})\, A \cdots B(Y_1)\, A\, B(Y_0)\, \pi(0)$$

for $k \geq 1$ and

$$v_{k,n} = M^{n-k} A^\top B(Y_{k+1}) A^\top B(Y_{k+2}) \cdots A' B(Y_n) \mathbf{1}$$

for $0 \leq k \leq n-1$.

Lemma 4.6 *For* $0 \leq k \leq n$

$$\overline{\mathbf{E}}\left[\Lambda_n X_k | \mathcal{Y}_n\right] = \operatorname{diag}\left(\langle q_k, e_i \rangle\right) v_{k,n} \,. \tag{4.2}$$

Proof The ith component of (4.2) is using the calculation for v above:

$$\begin{aligned}
&\overline{\mathbf{E}}\left[\Lambda_n \langle X_k, e_i \rangle \mid \mathcal{Y}_n\right] \\
&= \overline{\mathbf{E}}\left[\Lambda_k \langle X_k, e_i \rangle \overline{\mathbf{E}}\left[\Lambda_{k+1,n} \mid \mathcal{Y}_n \vee \{X_k = e_i\}\right] \mid \mathcal{Y}_n\right] \\
&= \overline{\mathbf{E}}\left[\Lambda_k \langle X_k, e_i \rangle v^i_{k,n} \mid \mathcal{Y}_n\right] \\
&= \overline{\mathbf{E}}\left[\Lambda_k \langle X_k, e_i \rangle \mid \mathcal{Y}_n\right] v^i_{k,n} \\
&= \langle q_k, e_i \rangle v^i_{k,n}
\end{aligned}$$

and the result follows. □

4.4 Exercises

Exercise 4.1 For the model

$$\begin{aligned}
X_{n+1} &= AX_n + V_{n+1} \\
Y_n &= CX_{n-1} + W_n
\end{aligned}$$

derive the analog of Theorem 4.4.

Exercise 4.2 For the model of Exercise 4.1 derive the analog of Theorem 4.5.

5
The Viterbi Algorithm

5.1 Introduction

This chapter discusses the most likely sequence of states of the hidden chain X which might have given rise to a given set of observations. This method is sometimes called decoding. It was introduced by Viterbi and has been widely employed in gene sequencing. The chapter concludes by describing how the Viterbi state estimates can be used to find the parameters of the model.

The goal is to find $i_0^*, i_1^*, \ldots, i_n^*$ so that given $\theta = (\pi(0), A, 0)$,

$$P_\theta(X_0 = e_{i_0^*}, Y_0 = y_0, \ldots, X_n = e_{i_n^*}, Y_n = y_n)$$

is maximized.

This is often called the decoding problem for a HMM and it could be solved using smoothers as was explained in Chapter 4. However, if the matrix A has some zero elements, then it may well be the case that the smoother solution will provide an unattainable solution as some of the transitions implied by this method may not be possible. There are many situations where elements of A are zero, for example if we use a higher-order chain model for X, or the states of the hidden chain have a meaning, (like credit ratings) where transitions are only possible to certain other states, (like nearby ratings).

The Viterbi method is used to deal with the case where it is possible that some elements of A are zero. When all elements of A are positive, then the Viterbi algorithm and the algorithm using smoothers agree.

Definition 5.1 For $k \geq 0$ and $1 \leq i \leq N$, define

$$\delta_k(i) = \max_{i_0, \ldots, i_{k-1}} P_\theta\left(Y_{0:k} = y_{0:k}, X_{0:k-1} = e_{i_{0:k-1}}, X_k = e_i\right).$$

This is the largest probability along a single path at time k which agrees with the $0:k$ observations and which ends in state e_i.

When $k=0$,

$$\delta_0(i) = P_\theta\left(Y_0 = y_0,\, X_0 = e_i\right).$$

Given the observations $y_{0:n}$, we write

$$\delta_n(i_n^*) = \max_{1\le i\le N} \delta_n(i).$$

Lemma 5.2 *For any $1 \le j \le N$ and $0 \le k \le n-1$,*

$$\delta_{k+1}(j) = B_j(y_{k+1}) \max_{1\le i\le N} \left[A_{ji}\, \delta_k(i)\right] \tag{5.1}$$

where

$$B_j(y) = \langle C e_j, y\rangle$$

and

$$\delta_0(j) = B_j(y_0)\, \pi_j(0). \tag{5.2}$$

Proof We first derive (5.2). In fact under $\theta = (\pi(0), A, C)$,

$$\begin{aligned}
\delta_0(j) &= P_\theta\left(Y_0 = y_0,\, X_0 = e_j\right)\\
&= P_\theta\left(Y_0 = y_0 \,|\, X_0 = e_j\right)\, P_\theta\left(X_0 = e_j\right) = \langle C e_j,\, y_0\rangle\, \pi_j(0).
\end{aligned}$$

To prove (5.1)

$$\begin{aligned}
&\delta_{k+1}(j)\\
&= \max_{i_0,\dots,i_k} \mathbf{E}_\theta\Big[\prod_{l=0}^{k}\langle X_l, e_{i_l}\rangle\,\langle X_{k+1}, e_j\rangle \prod_{l=0}^{k+1}\langle Y_l, y_l\rangle\Big]\\
&= \max_{i_0,\dots,i_k} \overline{\mathbf{E}}\Big[\overline{\Lambda}_{k+1}\prod_{l=0}^{k}\langle X_l, e_{i_l}\rangle\,\langle X_{k+1}, e_j\rangle \prod_{l=0}^{k+1}\langle Y_l, y_l\rangle\Big]\\
&= \max_{i_0,\dots,i_k} \overline{\mathbf{E}}\Big[\overline{\Lambda}_k MN\langle C X_{k+1}, Y_{k+1}\rangle\,\langle A X_k, X_{k+1}\rangle\\
&\qquad\qquad \prod_{l=0}^{k}\langle X_l, e_{i_l}\rangle\,\langle X_{k+1}, e_j\rangle \prod_{l=0}^{k+1}\langle Y_l, y_l\rangle\Big]\\
&= \langle C e_j, y_{k+1}\rangle \max_{i_0,\dots,i_k} \overline{\mathbf{E}}\Big[\overline{\Lambda}_k MN\langle A X_k, e_j\rangle \prod_{l=0}^{k}\langle X_l, e_{i_l}\rangle\\
&\qquad\qquad \prod_{l=0}^{k}\langle Y_l, y_l\rangle\,\langle X_{k+1}, e_j\rangle\,\langle Y_{k+1}, y_{k+1}\rangle\Big]
\end{aligned}$$

$$= \langle C e_j, y_{k+1} \rangle \max_{i_0,\ldots,i_k} \overline{\mathbf{E}}\left[\overline{\Lambda}_k MN\langle AX_k, e_j\rangle \prod_{l=0}^{k} \langle X_l, e_{i_l}\rangle \langle Y_l, y_l\rangle\right] \times$$

$$\overline{\mathbf{E}}\left[\langle X_{k+1}, e_j\rangle \langle Y_{k+1}, y_l\rangle\right]$$

$$= \langle C e_j, y_{k+1} \rangle \max_{i_0,\ldots,i_k} \overline{\mathbf{E}}\left[\overline{\Lambda}_k \langle AX_k, e_j\rangle \prod_{l=0}^{k} \langle X_l, e_{i_l}\rangle \langle Y_l, y_l\rangle\right]$$

$$= \langle C e_j, y_{k+1} \rangle \max_{i_0,\ldots,i_k} \overline{\mathbf{E}}\left[\overline{\Lambda}_k \langle AX_k, e_j\rangle \langle X_k, e_{i_k}\rangle \prod_{l=0}^{k-1} \langle X_l, e_{i_l}\rangle \prod_{l=0}^{k} \langle Y_l, y_l\rangle\right]$$

$$= \langle C e_j, y_{k+1} \rangle \max_{i_0,\ldots,i_k} \overline{\mathbf{E}}\left[\overline{\Lambda}_k \langle Ae_{i_k}, e_j\rangle \langle X_k, e_{i_k}\rangle \prod_{l=0}^{k-1} \langle X_l, e_{i_l}\rangle \prod_{l=0}^{k} \langle Y_l, y_l\rangle\right]$$

$$= \langle C e_j, y_{k+1} \rangle \max_{i_0,\ldots,i_k} \left[A_{ji_k} \overline{\mathbf{E}}\left[\overline{\Lambda}_k \langle X_k, e_{i_k}\rangle \prod_{l=0}^{k-1} \langle X_l, e_{i_l}\rangle \prod_{l=0}^{k} \langle Y_l, y_l\rangle\right]\right]$$

$$= B_j(y_{k+1}) \max_{1\le i_k \le N} \left[A_{ji_k} \max_{i_0,\ldots,i_{k-1}} \overline{\mathbf{E}}\left[\overline{\Lambda}_k \langle X_k, e_{i_k}\rangle \prod_{l=0}^{k} \langle X_l, e_{i_l}\rangle \langle Y_l, y_l\rangle\right]\right]$$

$$= B_j(y_{k+1}) \max_{1\le i \le N} \left[A_{ji} \max_{i_0,\ldots,i_{k-1}} \overline{\mathbf{E}}\left[\overline{\Lambda}_k \langle X_k, e_{i_k}\rangle \prod_{l=0}^{k} \langle X_l, e_{i_l}\rangle \langle Y_l, y_l\rangle\right]\right]$$

$$= B_j(y_{k+1}) \max_{1\le i \le N} \left[A_{ji} \max_{i_0,\ldots,i_{k-1}} \mathbf{E}_\theta\left[\langle X_k, e_{i_k}\rangle \prod_{l=0}^{k} \langle X_l, e_{i_l}\rangle \langle Y_l, y_l\rangle\right]\right]$$

$$= B_j(y_{k+1}) \max_{1\le i \le N} \left[A_{ji}\delta_k(i)\right].$$

This completes the proof of the lemma. □

Remark Rabiner and Juang (1993) also introduce

$$\psi_{k+1}(j) = \arg \max_{1\le i \le N} \left[A_{ji}\, \delta_k(i)\right].$$

If $i = \psi_{k+1}(j)$, then $A_{ji} > 0$. Such a choice will always be possible unless that jth row of A is zero.

5.2 Viterbi decoding

We first set

$$i_n^* = \arg \max_{1\le i \le N} \delta_n(i);$$

that is

$$\delta_n\left(i_n^*\right) = \max_{1\le i\le N} \delta_n(i).$$

Remark It is possible that there is more that one choice for i_n^* here. However, this is probably highly unlikely if the model is calibrated from experimental data. The same applies to i_k^* below.

Once i_n^* has been determined

$$\delta_n(i_n^*) = B_{i_n^*}\left(y_n\right) \max_{1\le i\le N} \left[A_{i_n^* i}\, \delta_{n-1}(i)\right]$$

and so

$$i_{n-1}^* = \arg \max_{1\le i\le N} \left[A_{i_n^* i}\, \delta_{n-1}(i)\right] = \psi_n\left(i_n^*\right).$$

In general for $0 \le k \le n-1$,

$$i_k^* = \psi_{k+1}\left(i_{k+1}^*\right).$$

Decoding with smoothers Recall that

$$P_\theta\left(X_k = e_i \,|\, Y_{0:n} = y_{0:n}\right) = \frac{\alpha_i(k)\,\beta_i(k)}{L_\theta}$$

where

$$L_\theta = \sum_{i=1}^{N} \alpha_i(k)\beta_i(k)$$

does not depend on k so for each $0 \le k \le n$ we find $i = i_k^*$ to maximize

$$\alpha_i(k)\,\beta_i(k).$$

For $k = 0$,

	$\alpha_i(0)$	$\beta_i(0)$	$\alpha_i(0)\,\beta_i(0)$
$i=1$	0.10	0.264	0.0264
$i=2$	0.03	0.246	0.0074
$i=3$	0.45	0.242	0.1089

so $i_0^* = 3$. For $k = 1$,

	$\alpha_i(1)$	$\beta_i(1)$	$\alpha_i(1)\,\beta_i(1)$
$i=1$	0.1080	0.46	0.0497
$i=2$	0.0177	0.50	0.0089
$i=3$	0.1683	0.50	0.0842

so $i_1^* = 3$. For $k = 2$,

	$\alpha_i(2)$	$\beta_i(2)$	$\alpha_i(2)\,\beta_i(2)$
$i = 1$	0.0516	1	0.0516
$i = 2$	0.0810	1	0.0810
$i = 3$	0.0102	1	0.0101

so $i_2^* = 2$.

Decoding with Viterbi For $k = 0$ and for $1 \leq i \leq 3$,

$$\delta_0(i) = \langle C\, e_i, y_0 \rangle\, \pi_i(0) = \left\langle C\, e_i, \begin{pmatrix} 1 \\ 0 \end{pmatrix} \right\rangle \pi_i(0) = B_i(y_0)\, \pi_i(0)$$

	$B_i(y_0)$	$\pi_i(0)$	$\delta_i(0)$
$i = 1$	0.5	0.2	0.10
$i = 2$	0.1	0.3	0.03
$i = 3$	0.9	0.5	0.45 .

For $k = 1$ and for $1 \leq j \leq 3$,

$$\delta_1(j) = B_j\,(y_1) \max_{1 \leq i \leq 3} \left[A_{ji}\, \delta_0(i) \right].$$

	$B_j(y_1)$	$A_{j1}\delta_0(1)$	$A_{j2}\delta_0(2)$	$A_{j3}\delta_0(3)$	$\psi_1(j)$	$\delta_1(j)$
$j = 1$	0.5	0.03	0.060	0.180	3	0.0900
$j = 2$	0.1	0.03	0.012	0.135	3	0.0135
$j = 3$	0.9	0.04	0.012	0.135	3	0.1215 .

For $k = 2$ and for $1 \leq j \leq 3$,

$$\delta_2(j) = B_j\,(y_2) \max_{1 \leq i \leq 3} \left[A_{ji}\, \delta_1(i) \right].$$

	$B_j(y_2)$	$A_{j1}\delta_1(1)$	$A_{j2}\delta_1(2)$	$A_{j3}\delta_1(3)$	$\psi_2(j)$	$\delta_2(j)$
$j = 1$	0.5	0.027	0.0027	0.04860	3	0.024300
$j = 2$	0.9	0.027	0.0054	0.03645	3	0.032905
$j = 3$	0.1	0.036	0.0054	0.03645	3	0.003645 .

So $i_2^* = 2$ which implies that $i_1^* = \psi_2(2) = 3$ and $i_0^* = \psi_1(3) = 3$. This is the same solution as was obtained using the smoothers.

5.3 Estimation of the model

Given $Y_{0:n} = y_{0:n}$, we shall determine $\theta = (\pi(0), A, C)$ to maximize the likelihood

$$P_\theta(Y_{0:n} = y_{0:n}) = L_\theta.$$

We shall employ the Expectation Maximization (EM) algorithm due to Baum and Welch. It has two steps: expectation followed by a maximization.

Lemma 5.3 *Given θ and $0 \le k \le n-1$ write*

$$\xi_k(j,i) = P_\theta\left(X_{k+1} = e_j \text{ and } X_k = e_i \,\middle|\, Y_{0:n} = y_{0:n}\right).$$

Then

$$\xi_k(j,i) = \frac{\alpha_k(i) A_{ji} B_j(y_{k+1}) \beta_{k+1}(j)}{L_\theta}$$

where

$$L_\theta = \sum_{i=1}^{N} \sum_{j=1}^{N} \alpha_k(i) A_{ji} B_j(y_{k+1}) \beta_{k+1}(j)$$

and

$$B_j(y) = \langle C\, e_j,\ y \rangle .$$

Proof For $0 \le k \le n-1$ and $i, j = 1, \ldots, n$,

$$\xi_k(j,i) = \frac{\mathbf{E}_\theta\left[\langle X_{k+1}, e_j\rangle \langle X_k, e_i\rangle \prod_{l=0}^{k} \langle Y_l, y_l\rangle\right]}{L_\theta}.$$

The numerator is

$$\mathbf{E}_\theta\left[\mathbf{E}_\theta\left[\langle X_{k+1}, e_j\rangle \langle X_k, e_i\rangle \prod_{l=0}^{k} \langle Y_l, y_l\rangle \,\middle|\, \mathcal{F}_n\right]\right]$$

$$= \mathbf{E}_\theta\left[\langle X_{k+1}, e_j\rangle \langle X_k, e_i\rangle\, \mathbf{E}_\theta\left[\prod_{l=0}^{k} \langle Y_l, y_l\rangle \,\middle|\, \mathcal{F}_n\right]\right]$$

$$= \mathbf{E}_\theta\left[\langle X_{k+1}, e_j\rangle \langle X_k, e_i\rangle \prod_{l=0}^{k} \langle C X_l, y_l\rangle\right]$$

$$= \langle C e_j, y_{k+1}\rangle\, \mathbf{E}_\theta\left[\langle X_k, e_i\rangle \prod_{l \ne k+1} \langle C X_l, y_l\rangle \,\middle|\, X_{k+1} = e_j\right] \times$$
$$P_\theta\left(X_{k+1} = e_j\right)$$

$$= \langle Ce_j, y_{k+1} \rangle \mathbf{E}_\theta \left[\langle X_k, e_i \rangle \prod_{l=0}^{k} \langle CX_l, y_l \rangle \middle| X_{k+1} = e_j \right] \times$$

$$\times \mathbf{E}_\theta \left[\prod_{l=k+2}^{n} \langle CX_l, y_l \rangle \middle| X_{k+1} = e_j \right] P_\theta (X_{k+1} = e_j) \quad \text{by Lemma 4.2}$$

$$= B_j(y_{k+1}) \mathbf{E}_\theta \left[\langle X_k, e_i \rangle \langle X_{k+1}, e_j \rangle \prod_{l=0}^{k} \langle CX_l, y_l \rangle \right] \times$$

$$\times \mathbf{E}_\theta \left[\prod_{l=k+2}^{n} \langle Y_l, y_l \rangle \middle| X_{k+1} = e_j \right]$$

$$= B_j(y_{k+1}) \beta_{k+1}(j) \mathbf{E}_\theta \left[\langle X_k, e_i \rangle \langle X_{k+1}, e_j \rangle \prod_{l=0}^{k} \langle CX_l, y_l \rangle \right]$$

$$= B_j(y_{k+1}) \beta_{k+1}(j) \mathbf{E}_\theta \left[\langle X_k, e_i \rangle \langle X_{k+1}, e_j \rangle \prod_{l=0}^{k} \langle CX_l, y_l \rangle \middle| X_k = e_i \right] \times$$

$$\times P_\theta (X_k = e_i)$$

$$= B_j(y_{k+1}) \beta_{k+1}(j) \mathbf{E}_\theta \left[\prod_{l=0}^{k} \langle CX_l, y_l \rangle \middle| X_k = e_i \right] \times$$

$$\times \mathbf{E}_\theta [\langle X_{k+1}, e_j \rangle \mid X_k = e_i] \; P_\theta (X_k = e_i)$$

$$= B_j(y_{k+1}) \beta_{k+1}(j) A_{ji} \mathbf{E}_\theta \left[\prod_{l=0}^{k} \langle CX_l, y_l \rangle \langle X_k = e_i \rangle \right]$$

$$= B_j(y_{k+1}) \beta_{k+1}(j) A_{ji} \mathbf{E}_\theta \left[\prod_{l=0}^{k} \langle Y_l, y_l \rangle \langle X_k = e_i \rangle \right]$$

$$= B_j(y_{k+1}) \beta_{k+1}(j) A_{ji} \alpha_i(k)$$

and the result follows. □

The EM algorithm says that if we suppose $\theta = (\pi(0), A, C)$ is an estimate for the parameters of a HMM, then an update with larger likelihood is $\hat{\theta} = (\hat{\pi}(0), \hat{A}, \hat{C})$ given by the formulae:

$$\hat{\pi}_i(0) = P_\theta (X_0 = e_i \mid Y_{0:n} = y_{0:n}) = \frac{\alpha_i(0)\beta_i(0)}{\sum_{i=1}^{N} \alpha_i(0)\beta_i(0)}$$

$$\hat{A}_{ji} = \frac{\sum_{k=0}^{n-1} \xi_k(j,i)}{\sum_{k=0}^{n-1} \gamma_k(i)} \quad \text{where} \quad \gamma_k(i) = \sum_{j=1}^{N} \xi_k(j,i)$$

and

$$\hat{C}_{ji} = \frac{\sum_{k=0}^{n} \langle y_k, f_j \rangle \gamma_k(i)}{\sum_{k=0}^{n} \gamma_k(i)}.$$

These updates can now be used and the estimation be iterated. The justification of this procedure follows from the EM algorithm which is discussed in the next section.

The size of the state space of the HMM, the number N, can also be estimated, but usually a practitioner will try some reasonable choices.

5.4 Exercises

Exercise 5.1 (See Rabiner and Juang, 1993, page 341.) Let $N = 3, M = 2$. Then

$$\theta = \left(\pi(0) = \frac{1}{3} \begin{bmatrix} 1 \\ 1 \\ 1 \end{bmatrix}, A = \frac{1}{3} \begin{bmatrix} 1 & 1 & 1 \\ 1 & 1 & 1 \\ 1 & 1 & 1 \end{bmatrix}, C = \begin{bmatrix} 0.5 & 0.75 & 0.25 \\ 0.5 & 0.25 & 0.75 \end{bmatrix} \right).$$

The observations are

$$f_1 = \begin{pmatrix} 1 \\ 0 \end{pmatrix}, f_2 = \begin{pmatrix} 0 \\ 1 \end{pmatrix}.$$

phase	1	2	3
$P(f_1)$	0.5	0.75	0.25
$P(f_2)$	0.5	0.25	0.75

If the observation is HHHHTHTTTT, show that the decoding is 2222323333.

For a simpler problem, if the observation is HHT, show the decoding is 223.

Exercise 5.2 Let $M = 2, N = 3$ and $n = 2$. Then

$$\pi(0) = \begin{bmatrix} 0.2 \\ 0.3 \\ 0.5 \end{bmatrix}, A = \begin{bmatrix} 0.3 & 0.2 & 0.4 \\ 0.3 & 0.4 & 0.3 \\ 0.4 & 0.4 & 0.3 \end{bmatrix}, C = \begin{bmatrix} 0.5 & 0.1 & 0.9 \\ 0.5 & 0.9 & 0.1 \end{bmatrix}$$

The observation is f, f_1, f_2 (see Example 3.10). Decode.

6
The EM Algorithm

6.1 Introduction

This chapter introduces a recursive procedure which estimates the parameters of the model; that is, the transition probabilities of the hidden chain, the transition probabilities of the observed process Y, which depend on the state of the hidden chain, and the initial distribution of the chain.

That is, we seek $\theta = (\pi(0), A, C)$ so that, given observations

$$y_{0:n} = (y_0, y_1, \dots, y_n),$$

the probability

$$P_\theta \left(Y_{0:n} = y_{0:n} \right)$$

is a maximum.

In other words, we seek a maximum likelihood estimator for θ.

This problem is rarely solvable exactly and the EM algorithm provides a sequence $\{\hat{\theta}_p\}$ of parameters that provide increasing values for the likelihood.

This is one of the usual methods of estimating the parameters in the HMM model. It is also used widely in statistics to maximimize likelihood in different contexts as discussed in Elliott et al. (1995). The books McLachlan and Peel (2000), McLachlan and Krishnan (2008) give more general applications, as does Dempster et al. (1977), pp. 1-22.

Lemma 6.1 *Given* $\theta = (\pi(0), A, C)$,

$$P_\theta \left(Y_{0:n} = y_{0:n} \right) = \frac{1}{M^{n+1}} \overline{\mathbf{E}} \left[\overline{\Lambda}_n(\theta) \,\middle|\, Y_{0:n} = y_{0:n} \right],$$

where

$$\overline{\Lambda}_n(\theta) = M^{n+1} N^{n+1} \langle \pi(0), X_0 \rangle \prod_{l=1}^{n} \langle A X_{l-1}, X_l \rangle \prod_{l=0}^{n} \langle C X_l, Y_l \rangle.$$

Proof We have

$$\begin{aligned}
P_\theta \left(Y_{0:n} = y_{0:n} \right) &= \mathbf{E}_\theta \left[\prod_{l=0}^{n} \langle Y_l, y_l \rangle \right] \\
&= \overline{\mathbf{E}} \left[\overline{\Lambda}_n(\theta) \prod_{l=0}^{n} \langle Y_l, y_l \rangle \right] \\
&= \overline{\mathbf{E}} \left[\overline{\mathbf{E}} \left[\overline{\Lambda}_n(\theta) \prod_{l=0}^{n} \langle Y_l, y_l \rangle \,\middle|\, Y_{0:n} \right] \right] \\
&= \overline{\mathbf{E}} \left[\prod_{l=0}^{n} \langle Y_l, y_l \rangle \, \overline{\mathbf{E}} \left[\overline{\Lambda}_n(\theta) \,\middle|\, Y_{0:n} \right] \right] \\
&= \overline{\mathbf{E}} \left[\prod_{l=0}^{n} \langle Y_l, y_l \rangle \, \overline{\mathbf{E}} \left[\overline{\Lambda}_n(\theta) \,\middle|\, Y_{0:n} = y_{0:n} \right] \right] \\
&= \overline{\mathbf{E}} \left[\prod_{l=0}^{n} \langle Y_l, y_l \rangle \right] \overline{\mathbf{E}} \left[\overline{\Lambda}_n(\theta) \,\middle|\, Y_{0:n} = y_{0:n} \right] \\
&= \frac{1}{M^{n+1}} \overline{\mathbf{E}} \left[\overline{\Lambda}_n(\theta) \,\middle|\, Y_{0:n} = y_{0:n} \right]
\end{aligned}$$

and we are done. □

Lemma 6.2 *Let*

$$Q(\theta', \theta) \equiv \mathbf{E}_\theta \left[\log \frac{\overline{\Lambda}_n(\theta')}{\overline{\Lambda}_n(\theta)} \,\middle|\, Y_{0:n} = y_{0:n} \right],$$

then

$$\log L_{\theta'} - \log L_\theta \geq Q(\theta', \theta),$$

where $L_\theta \equiv P_\theta \left(Y_{0:n} = y_{0:n} \right)$ *is the likelihood with parameter* θ.

Remark The prime indicates a different choice of parameter. Thus, $\theta' = (\pi'(0), A', C')$ is another possible choice of parameters. The function Q is called a *pseudo-log-likelihood.*

Proof We can use Jensen's inequality (see Cohen and Elliott, 2015, Lemma 1.9)

$$\varphi \left(\mathbf{E} \left[X \,|\, \mathcal{A} \right] \right) \leq \mathbf{E} \left[\varphi(X) \,|\, \mathcal{A} \right]$$

if φ is convex. We use $\varphi(x) = -\log x$ where $x > 0$, and note that

$$\varphi'(x) = -\frac{1}{x} \text{ and } \varphi''(x) = \frac{1}{x^2} > 0.$$

This implies that

$$-\log \mathbf{E}[X \mid \mathcal{A}] \leq \mathbf{E}[-\log X \mid \mathcal{A}] \quad \text{if } x > 0,$$

so

$$\mathbf{E}[\log X \mid \mathcal{A}] \leq \log \mathbf{E}[X \mid \mathcal{A}].$$

Thus, using Bayes' identity, Appendix C,

$$\begin{aligned}
Q(\theta', \theta) &= \mathbf{E}_\theta\left[\log \frac{\overline{\Lambda}_n(\theta')}{\overline{\Lambda}_n(\theta)} \,\middle|\, Y_{0:n} = y_{0:n}\right] \\
&\leq \log \mathbf{E}_\theta\left[\frac{\overline{\Lambda}_n(\theta')}{\overline{\Lambda}_n(\theta)} \,\middle|\, Y_{0:n} = y_{0:n}\right] \\
&= \log \frac{\overline{\mathbf{E}}\left[\overline{\Lambda}_n(\theta)\left\{\frac{\overline{\Lambda}_n(\theta')}{\overline{\Lambda}_n(\theta)}\right\} \,\middle|\, Y_{0:n} = y_{0:n}\right]}{\overline{\mathbf{E}}\left[\overline{\Lambda}_n(\theta) \,\middle|\, Y_{0:n} = y_{0:n}\right]} \\
&= \log \frac{\overline{\mathbf{E}}\left[\overline{\Lambda}_n(\theta') \,\middle|\, Y_{0:n} = y_{0:n}\right]}{\overline{\mathbf{E}}\left[\overline{\Lambda}_n(\theta) \,\middle|\, Y_{0:n} = y_{0:n}\right]} = \log\left[\frac{L_{\theta'}}{L_\theta}\right] = \log L_{\theta'} - \log L_\theta,
\end{aligned}$$

and we are done. □

6.2 Steps of the EM algorithm

Step 1: Choose an initial value $\theta = \hat{\theta}_0$.

This could be made by choosing a previous estimate, or we could choose $\hat{\theta}_0 = \bar{\theta} = (\overline{\pi}(0), \overline{A}, \overline{C})$ where

$$\overline{\pi}(0) = \frac{1}{N}\begin{bmatrix}1\\1\\\vdots\\1\end{bmatrix} \equiv \frac{1}{N}\mathbf{1}_N, \ \overline{A} = \frac{1}{N}\begin{bmatrix}1 & 1 & \dots & 1\\1 & 1 & \dots & 1\\\vdots & \vdots & \ddots & \vdots\\1 & 1 & \dots & 1\end{bmatrix},$$

and

$$\overline{C} = \frac{1}{M}\begin{bmatrix}1 & 1 & \dots & 1\\1 & 1 & \dots & 1\\\vdots & \vdots & \ddots & \vdots\\1 & 1 & \dots & 1\end{bmatrix}$$

for then

$$\overline{\Lambda}_n\left(\overline{\theta}\right) \equiv 1.$$

This follows because

$$\begin{aligned}
\langle \overline{\pi}(0), X_0 \rangle &= \frac{1}{N}\langle \mathbf{1}, X_0 \rangle = \frac{1}{N}\,, \\
\langle \overline{A} X_{l-1}, X_l \rangle &= \left\langle \frac{1}{N}\mathbf{1}_N, X_l \right\rangle = \frac{1}{N}\,, \\
\langle \overline{C} X_l, Y_l \rangle &= \left\langle \frac{1}{M}\mathbf{1}_M, Y_l \right\rangle = \frac{1}{M}\,.
\end{aligned}$$

Step 2 (The Expectation, E, step): Given $\theta = \hat{\theta}_p$, compute $Q(\theta', \theta)$.

.

Step 3 (The maximization, M, step): Maximize $Q(\theta', \theta)$ with respect to θ' and to obtain $\hat{\theta}_{p+1}$.

We note that $\theta' = \theta$ implies $Q(\theta', \theta) = 0$ and so $Q(\hat{\theta}_{p+1}, \theta) \geq 0$. Usually, $Q(\theta', \theta) > 0$ with $\theta' = \hat{\theta}_{p+1}$ and this will imply $L_{\theta'} > L_{\theta}$.

Step 4: Replace $\hat{\theta}_p$ with $\hat{\theta}_{p+1}$ and return to Step 2.

The expectatation, E, step

We need only maximize

$$\mathbf{E}_\theta\left[\log \overline{\Lambda}_n(\theta') \,\middle|\, Y_{0:n} = y_{0:n}\right]$$

because

$$Q(\theta', \theta) = \mathbf{E}_\theta\left[\log \overline{\Lambda}_n(\theta') \,\middle|\, Y_{0:n} = y_{0:n}\right] - \mathbf{E}_\theta\left[\log \overline{\Lambda}_n(\theta) \,\middle|\, Y_{0:n} = y_{0:n}\right].$$

We also note that

$$\overline{\Lambda}_n(\theta') = \text{constant} \cdot \overline{\Lambda}_n^{\pi}(\theta') \cdot \overline{\Lambda}_n^{A}(\theta') \cdot \overline{\Lambda}_n^{C}(\theta'),$$

where

$$\begin{aligned}
\text{constant} &= M^{n+1} N^{n+1}, \\
\overline{\Lambda}_n^{\pi}(\theta') &= \langle \pi'(0), X_0 \rangle, \\
\overline{\Lambda}_n^{A}(\theta') &= \prod_{l=1}^{n} \langle A' X_{l-1}, X_l \rangle, \\
\overline{\Lambda}_n^{C}(\theta') &= \prod_{l=0}^{n} \langle C' X_l, Y_l \rangle.
\end{aligned}$$

Then

$$\begin{aligned}
&\mathbf{E}_\theta\left[\log \overline{\Lambda}_n(\theta') \,\middle|\, Y_{0:n} = y_{0:n}\right] \\
&\quad = \text{constant} + \mathbf{E}_\theta\left[\log \overline{\Lambda}_n^\pi(\theta') \,\middle|\, Y_{0:n} = y_{0:n}\right] + \\
&\qquad + \mathbf{E}_\theta\left[\log \overline{\Lambda}_n^A(\theta') \,\middle|\, Y_{0:n} = y_{0:n}\right] + \mathbf{E}_\theta\left[\log \overline{\Lambda}_n^C(\theta') \,\middle|\, Y_{0:n} = y_{0:n}\right].
\end{aligned}$$

We now compute each of the three terms on the right-hand side of this expression. Set

$$\begin{aligned}
T_1(\theta') &= \mathbf{E}_\theta\left[\log\langle \pi'(0), X_0\rangle \,\middle|\, Y_{0:n} = y_{0:n}\right] \\
T_2(\theta') &= \sum_{l=1}^{n} \mathbf{E}_\theta\left[\log\langle A' X_{l-1}, X_l\rangle \,\middle|\, Y_{0:n} = y_{0:n}\right] \\
T_3(\theta') &= \sum_{l=0}^{n} \mathbf{E}_\theta\left[\log\langle C' X_l, Y_l\rangle \,\middle|\, Y_{0:n} = y_{0:n}\right].
\end{aligned}$$

Formula for $T_1(\theta')$. We have

$$\begin{aligned}
T_1(\theta') &= \mathbf{E}_\theta\left[\sum_{i=1}^{N} \langle X_0, e_i\rangle \log\langle \pi'(0), X_0\rangle \,\middle|\, Y_{0:n} = y_{0:n}\right] \\
&= \sum_{i=1}^{N} \log \pi_i'(0)\, \mathbf{E}_\theta\left[\langle X_0, e_i\rangle \,\middle|\, Y_{0:n} = y_{0:n}\right] \\
&= \sum_{i=1}^{N} \gamma_i(0) \log \pi_i'(0),
\end{aligned}$$

where

$$\gamma_i(0) = \frac{\alpha_i(0)\beta_i(0)}{\sum_{i=1}^{N} \alpha_i(0)\beta_i(0)} = \frac{\alpha_i(0)\beta_i(0)}{L_\theta}$$

using smoothers.

Formula for $T_2(\theta')$. We also have

$$\begin{aligned}
T_2(\theta') &= \sum_{l=1}^{n}\sum_{i=1}^{N}\sum_{j=1}^{N} \log A_{ji}'\, \mathbf{E}_\theta\left[\langle X_l, e_j\rangle\langle X_{l-1}, e_i\rangle \,\middle|\, Y_{0:n} = y_{0:n}\right] \\
&= \sum_{l=1}^{n}\sum_{i=1}^{N}\sum_{j=1}^{N} \xi_{l-1}(j,i) \log A_{ji}'.
\end{aligned}$$

Formula for $T_3(\theta')$. Finally,

$$\begin{aligned} T_3(\theta') &= \sum_{l=0}^{n}\sum_{i=1}^{N}\sum_{j=1}^{M} \log C'_{ji}\, \mathbf{E}_\theta\left[\langle X_l, e_i\rangle\langle Y_l, f_j\rangle \mid Y_{0:n} = y_{0:n}\right] \\ &= \sum_{l=0}^{n}\sum_{i=1}^{N}\sum_{j=1}^{M} \log C'_{ji}\, \langle y_l, f_j\rangle\, \mathbf{E}_\theta\left[\langle X_l, e_i\rangle \mid Y_{0:n} = y_{0:n}\right] \\ &= \sum_{l=0}^{n}\sum_{i=1}^{N}\sum_{j=1}^{M} \langle y_l, f_j\rangle\, \gamma_l(i)\, \log C'_{ji}. \end{aligned}$$

The **estimation** step is completed.

The maximization, M, step

We proceed to the **maximization** stage.

Lemma 6.3 *Let $a_1, a_2, \ldots, a_n \geq 0$ with*

$$A = \sum_{i=1}^{n} a_i > 0.$$

Let

$$\mathcal{U} = \left\{ \mathbf{x} \in \mathbb{R}^n \,\middle|\, x_i > 0 \text{ for all } i \text{ with } \sum_{i=1}^{n} x_i = 1 \right\}.$$

Then the supremum of

$$f(\mathbf{x}) = \sum_{i=1}^{n} a_i \log x_i, \quad \mathbf{x} \in \mathcal{U}$$

is attained at $\mathbf{x} = \hat{\mathbf{x}}$, where

$$\hat{x}_i = \begin{cases} 0 & \text{if } a_i = 0 \\ a_i/A & \text{if } a_i > 0. \end{cases}$$

Remark Note that $f(\mathbf{x}) \leq 0$ for $\mathbf{x} \in \mathcal{U}$.

If $a_i = 0$ for some i, then the supremum of f on $\mathcal{U}$ is not attained on $\mathcal{U}$.

Proof If $a, b > 0$, then as in Lemma 2.2

$$\log a \leq \log b + \frac{1}{b}(a - b).$$

Let $\mathbf{x} \in \mathcal{U}$. We have

$$\log x_i \leq \log \hat{x}_i + \frac{1}{\hat{x}_i}(x_i - \hat{x}_i) \text{ if } a_i > 0.$$

This implies that

$$\begin{aligned} f(\mathbf{x}) &= \sum_{i=1, a_i>0}^{n} a_i \log x_i \\ &\leq \sum_{i=1, a_i>0}^{n} a_i \left[\log \hat{x}_i + \frac{1}{\hat{x}_i}(x_i - \hat{x}_i)\right] \\ &= f(\hat{\mathbf{x}}) + \sum_{i=1, a_i>0}^{n} a_i \frac{1}{\hat{x}_i}(x_i - \hat{x}_i) \\ &= f(\hat{\mathbf{x}}) + \sum_{i=1, a_i>0}^{n} a_i \frac{A}{a_i}(x_i - \hat{x}_i) \\ &= f(\hat{\mathbf{x}}) + A \cdot \sum_{i=1}^{n}(x_i - \hat{x}_i) \\ &= f(\hat{\mathbf{x}}). \end{aligned}$$

and we are done. □

Maximization of T_1. Recall that

$$T_1(\theta') = \sum_{i=1}^{N} \gamma_i(0) \log \pi'_i(0).$$

As above the maximum is given by

$$\pi'_i(0) = \begin{cases} \gamma_i(0)/\sum_{i=1}^{N} \gamma_i(0) & \text{if } \gamma_i(0) > 0 \\ 0 & \text{otherwise.} \end{cases}$$

But

$$\sum_{i=1}^{N} \gamma_i(0) = \mathbf{E}_\theta \left[\langle X_0, e_i \rangle \,\middle|\, Y_{0:n} = y_{0:n}\right] = 1.$$

So $\hat{\pi}_i(0) = \gamma_i(0)$ for $i = 1, \ldots, N$, and the update is given by

$$\hat{\pi}(0) = \gamma(0).$$

Maximization of T_2. Recall that

$$\begin{aligned} T_2(\theta') &= \sum_{l=1}^{n}\sum_{i=1}^{N}\sum_{j=1}^{N} \xi_{l-1}(j,i)\,\log A'_{ji} \\ &= \sum_{i=1}^{N}\left\{\sum_{j=1}^{N}\left(\sum_{l=1}^{n}\xi_{l-1}(j,i)\right)\log A'_{ji}\right\}. \end{aligned}$$

Recall also that $\gamma_{l-1}(i) = \sum_{j=1}^{N} \xi_{l-1}(j,i)$. If we assume that

$$\sum_{j=1}^{N}\sum_{l=1}^{n}\xi_{l-1}(j,i) = \sum_{l=1}^{n}\gamma_{l-1}(i) = \sum_{l=0}^{n-1}\gamma_l(i) > 0,$$

then for such i, the update – that is, the maximizer for A_{ji} – is again given by

$$\hat{A}_{ji} = \frac{\sum_{l=1}^{n}\xi_{l-1}(j,i)}{\sum_{j=1}^{N}\sum_{l=1}^{n}\xi_{l-1}(j,i)} = \frac{\sum_{l=0}^{n-1}\xi_l(j,i)}{\sum_{l=0}^{n-1}\gamma_l(i)}.$$

However if

$$\sum_{l=0}^{n-1}\gamma_l(i) = 0$$

we have no new information to update A_{ji} and so we retain the current value of A_{ji} for that value of i.

Maximization for T_3. Recall that

$$\begin{aligned} T_3(\theta') &= \sum_{l=0}^{n}\sum_{i=1}^{N}\sum_{j=1}^{M} \gamma_l(i)\,\langle y_l, f_j\rangle\,\log C'_{ji} \\ &= \sum_{i=1}^{N}\left\{\sum_{j=1}^{M}\left(\sum_{l=0}^{n}\gamma_l(i)\,\langle y_l, f_j\rangle\right)\log C'_{ji}\right\}. \end{aligned}$$

We note that

$$\sum_{j=1}^{M}\sum_{l=0}^{n}\gamma_l(i)\,\langle y_l, f_j\rangle = \sum_{l=0}^{n}\gamma_l(i)$$

so if

$$\sum_{l=0}^{n}\gamma_l(i) > 0$$

then again the update is given by

$$\hat{C}_{ji} = \frac{\sum_{l=0}^{n} \langle y_l, f_j \rangle \, \gamma_l(i)}{\sum_{l=0}^{n} \gamma_l(i)} .$$

If

$$\sum_{l=0}^{n} \gamma_l(i) = 0$$

we do not update C_{ji} for that value of i.

Example 6.4 Let $M = 2$, $N = 3$ and $n = 2$.

$$\pi(0) = \begin{bmatrix} 0.2 \\ 0.3 \\ 0.5 \end{bmatrix}, \quad A = \begin{bmatrix} 0.3 & 0.2 & 0.4 \\ 0.3 & 0.4 & 0.3 \\ 0.4 & 0.4 & 0.3 \end{bmatrix}, \quad C = \begin{bmatrix} 0.5 & 0.75 & 0.25 \\ 0.25 & 0.25 & 0.75 \end{bmatrix}$$

The observation is $f_1 f_1 f_2$.

We then have

$$\hat{\pi}(0) = \gamma(0) = \begin{bmatrix} 0.1850 \\ 0.0517 \\ 0.7632 \end{bmatrix} .$$

With $L = 0.1427$, we have the expression

$$\gamma_i(0) = \frac{\alpha_i(0)\beta_i(0)}{L}$$

which can be written in matrix form as

$$\gamma(0) = \frac{1}{L} \, \mathtt{diag}\big(\alpha(0)\big) * \beta(0),$$

where again we have used MATLAB. Similarly,

$$\xi_k = \frac{1}{L} \left(B\left(y_{k+1}\right) * \mathtt{diag}\big(\beta(k+1)\big) * A * \mathtt{diag}\big(\alpha(K)\big)\right)$$

and

$$\gamma_k = \mathtt{transpose}(\xi_k) * \mathbf{1} \quad \text{for } k \geq 1.$$

In this example

$$\sum_{l=0}^{n-1} \gamma_l(i) > 0$$

for each i and so the update for A is

$$\hat{A}_{ji} = \frac{\sum_{l=0}^{n-1} \xi_l(j,i)}{\sum_{l=0}^{n-1} \gamma_l(i)} .$$

This can be written in matrix form as

$$\hat{A} = \left(\sum_{l=0}^{n-1} \xi_l\right) * \texttt{inv}\left(\texttt{diag}\left(\sum_{l=0}^{n-1} \gamma_l\right)\right).$$

For this example, we have

$$\hat{A} = \begin{bmatrix} 0.3036 & 0.1941 & 0.3889 \\ 0.4030 & 0.4296 & 0.2703 \\ 0.2934 & 0.3763 & 0.3408 \end{bmatrix}.$$

The update for C is now calculated.

Let

$$Y = [\, y_0 \; y_1 \; y_2 \,] = \begin{bmatrix} 1 & 1 & 0 \\ 0 & 0 & 1 \end{bmatrix}$$

and

$$G = [\, \gamma_0 \; \gamma_1 \; \cdots \; \gamma_n \,]$$

Then the (j, i)th element of

$$Y * \texttt{transpose}\,(G)$$

is

$$\sum_{l=0}^{n} \langle y_l, f_j \rangle \, \gamma_l(i)$$

as

$$Y_{j\,l} = \langle y_l, f_j \rangle$$

and

$$\texttt{transpose}\,(G)_{l\,i} = G_{i\,l} = \gamma_l(i)$$

and so

$$\hat{C} = Y * \texttt{transpose}(G) * \texttt{inv}\left[\texttt{diag}\left(\sum_{l=0}^{n} \gamma_l\right)\right].$$

The update for C in this example is

$$\hat{C} = \begin{bmatrix} 0.5957 & 0.1670 & 0.9504 \\ 0.4043 & 0.8330 & 0.0496 \end{bmatrix}.$$

6.3 Exercises

Exercise 6.1 Check the stages in the maximization step for T_2.

Exercise 6.2 Check the stages in the maximization step for T_3.

Exercise 6.3 Rework Example 6.4 with the data of Exercise 3.2.

7

A New Markov Chain Model

7.1 Introduction

A modified hidden Markov model is discussed in this chapter. The hidden process X is a finite state Markov chain as before. However, the observed process Y is now itself a finite state Markov chain whose transition probabilities depend on the state of the hidden process X. The chapter first constructs these processes on the canonical probability space. The results of previous chapters are then developed for the new model. Filters and smoothers are derived, as well as the Viterbi estimates. The EM algorithm is then applied to the new model.

For the standard hidden Markov model, we have the semi-martingale representation

$$X_{k+1} = AX_k + M_{k+1}$$

for the hidden chain and for the observed chain

$$Y_k = C\,X_k + N_k$$

with parameters $\theta = (\pi(0), A, C)$.

We have seen that, given the hidden chain, the terms of the observed chain are independent: see Lemma 3.7. This property may not be suitable for DNA modeling so we now give a model which does not have this restriction.

For the observation chain we now postulate that

$$Y_k = C(X_k)Y_{k-1} + N_k\,.$$

This means that $\{Y_k\}$ is a first-order Markov chain but its transition matrix depends on the phase (state) of the hidden chain X. The model is introduced in van der Hoek and Elliott (2013). The calculations are similar to those in earlier chapters so some details are omitted.

We now require that

$$P(Y_k = f_j \mid \mathcal{F}_k \vee \mathcal{Y}_{k-1}) = P(Y_k = f_r \mid X_k \vee Y_{k-1})$$

and more specifically,

$$P(Y_k = f_r \mid X_k = e_i, Y_{k-1} = f_s) = C^i_{rs}$$

where

$$C(X_k) = \sum_{i=1}^{N} C^i \langle X_k, e_i \rangle .$$

We shall now present all the algorithms for this model.

7.2 Construction of the model

In addition to the parameters $(\pi(0), A, C)$, we shall also consider the $M \times N$ matrix

$$D = (D_{ri},\, 1 \le r \le M. 1 \le i \le N)$$

where

$$D_{ri} = P(Y_0 = f_r | X_0 = e_i).$$

However, it is difficult to estimate D without multiple observation chains.

Write $\theta = (\pi(0), A, C, D)$ for the parameters of this model. Here D is as above, $\pi(0)$ is the initial distribution of X, $A = (A_{ji}\,,\, 1 \le i, j \le N)$ where

$$A_{ji} = P(X_{k+1} = e_j \mid X_k = e_i)$$

and $C = (C^i\,,\, 1 \le i \le N)$ is a set of transition matrices for Y. Note we assume the parameters A_{ji}, C^i_{sr} do not depend on the index k. For the case where they do depend on k, some parametric specification would be required.

Reference probability The dynamics described above are those under the 'real world' probability P. To facilitate the calculations as before we introduce a 'reference probability' $\overline{P}$ under which at each time k the values of X_k and Y_k are iid random variables taking the values in $\{e_1, \cdots e_N\}$ and $\{f_1, \cdots f_M\}$ respectively.

That is, under $\overline{P}$,

$$\overline{P}(X_k = e_i) = \frac{1}{N}, \qquad \overline{P}(Y_k = f_r) = \frac{1}{M}$$
$$\overline{P}(X_k = e_i \text{ and } Y_k = f_r) = \frac{1}{MN}.$$

A sample path of the two processes (X, Y) to time k is a sequence of $k+1$ values (e_i, f_j). The probability of that sample path under $\overline{P}$ is

$$\frac{1}{(MN)^{k+1}}.$$

Given the set of parameters θ, the 'real world' measure P is now defined in terms of $\overline{P}$.

For $l = 0$, define

$$\overline{\lambda}_0 = NM \langle \pi(0), X_0 \rangle \langle DX_0, Y_0 \rangle .$$

Write

$$C(X_l) = \sum_{i=1}^{N} C^i \langle X_l, e_i \rangle .$$

For $l \geq 1$, define

$$\overline{\lambda}_l = MN \langle AX_{l-1}, X_l \rangle \langle C(X_l) Y_{l-1}, Y_l \rangle .$$

Then, for $k \geq 0$, write

$$\overline{\Lambda}_k = \prod_{l=0}^{k} \overline{\lambda}_l .$$

Lemma 7.1 *For each $l \geq 1$,*

$$\overline{\mathbf{E}}\left[\overline{\lambda}_l \mid X_{0:l-1} \vee Y_{0:l-1}\right] = 1$$

and

$$\overline{\mathbf{E}}[\overline{\lambda}_0] = 1.$$

Proof Let $l \geq 1$.

$$\overline{\mathbf{E}}\left[NM\langle C(X_l)Y_{l-1}, Y_l\rangle\langle AX_{l-1}, X_l\rangle \mid X_{0:l-1} \vee Y_{0:l-1}\right]$$
$$= \overline{\mathbf{E}}\Bigg[\sum_{i=1}^{N}\sum_{j=1}^{M}\langle X_l, e_i\rangle\langle Y_l, f_j\rangle NM\langle C(X_l)Y_{l-1}, Y_l\rangle \times$$
$$\times \langle AX_{l-1}, X_l\rangle \,\Big|\, X_{0:l-1} \vee Y_{0:l-1}\Bigg]$$

$$= NM\sum_{i=1}^{N}\sum_{j=1}^{M}\langle Ce_i, f_j\rangle\langle AX_{l-1}, e_i\rangle\,\overline{\mathbf{E}}[\langle X_l, e_i\rangle\langle Y_l, f_j\rangle \mid X_{0:\,l-1}\vee Y_{0:\,l-1}]$$

$$= \sum_{i=1}^{N}\langle AX_{l-1}, e_i\rangle \quad \text{as} \quad \sum_{j=1}^{M} C_{ji} = 1$$

$$= \langle AX_{l-1}, 1\rangle$$

$$= 1 \quad \text{as } \langle AX_{l-1}, 1\rangle = \mathbf{1}^\top AX_{l-1} = \mathbf{1}^\top X_{l-1} = 1.$$

The second identity is proved in a similar way. □

The relationship between $\overline{P}$ and $P = P_\theta$ is then given by noting that

$$\left.\frac{dP_\theta}{d\overline{P}}\right|_{X_{0:\,k}\vee Y_{0:\,k}} = \overline{\Lambda}_k$$

for each $k = 0, 1, 2, \ldots$.

Of course $\overline{P}$ can be expressed in terms of $P = P_\theta$ by

$$\left.\frac{d\bar{P}}{dP_\theta}\right|_{X_{0:k}\vee Y_{0:k}} = \left(\bar{\Lambda}_k\right)^{-1}.$$

Similarly to the results in Elliott et al. (1995), we now prove that with the probability P expressed this way, $\{X_k, Y_k\}$ will have the desired dynamics with the given parameters. That is, under P,

$$P(X_0 = e_i) = \pi_i(0)$$

for each $1 \le i \le N$,

$$P(X_{k+1} = e_j \mid X_k = e_i) = A_{ji}$$

for each $1 \le i, j \le N$, and

$$P(Y_k = f_s \mid X_k = e_i, Y_{k-1} = f_r) = C^i_{sr}$$

for each $1 \le r, s \le M$ and $1 \le i \le N$.

These facts can be proved as in Elliott et al. (1995), so we omit the details except for the final identity.

Lemma 7.2 *For $1 \le r, s \le M$ and $1 \le i \le N$, we have*

$$P(Y_k = f_s \mid X_k = e_i, Y_{k-1} = f_r) = C^i_{sr}.$$

In fact we shall prove

$$P(Y_k = f_s \mid Y_{0:k-1} \vee X_{0:k}) = \langle C(X_k)\, Y_{k-1},\, f_s\rangle$$

from which the first identity follows by further conditioning.

Proof In fact

$$
\begin{aligned}
P(Y_k = f_s \mid Y_{0:k-1} \vee X_{0:k}) \\
&= \mathbf{E}\left[\langle Y_k, f_s\rangle \mid Y_{0:k-1} \vee X_{0:k}\right] \\
&= \frac{\overline{\mathbf{E}}\left[\overline{\Lambda}_k \langle Y_k, f_s\rangle \mid Y_{0:k-1} \vee X_{0:k}\right]}{\overline{\mathbf{E}}\left[\overline{\Lambda}_k \mid Y_{0:k-1} \vee X_{0:k}\right]} \quad \text{by Bayes' Theorem} \\
&= \frac{\overline{\mathbf{E}}\left[\langle C(X_k)\, Y_{k-1},\, Y_k\rangle \langle Y_k, f_s\rangle \mid Y_{0:k-1} \vee X_{0:k}\right]}{\overline{\mathbf{E}}\left[\langle C(X_k)\, Y_{k-1},\, Y_k\rangle \mid Y_{0:k-1} \vee X_{0:k}\right]}.
\end{aligned}
$$

Here the denominator is

$$
\begin{aligned}
&\overline{\mathbf{E}}\left[\langle C(X_k)\, Y_{k-1},\, Y_k\rangle \mid \mathcal{G}_{k-1} \vee X_k\right] \\
&= \sum_{s=1}^{M} \overline{\mathbf{E}}\left[\langle Y_k,\, f_s\rangle \langle C(X_k)\, Y_{k-1},\, Y_k\rangle \mid \mathcal{G}_{k-1} \vee X_k\right] \\
&= \frac{1}{M} \mathbf{1}^\top C(X_k)\, Y_{k-1} = \frac{1}{M} \mathbf{1}^\top Y_{k-1} = \frac{1}{M}
\end{aligned}
$$

since the column sums of $C(X_k)$ are 1 for each choice of X_k. Thus,

$$
\begin{aligned}
&\overline{\mathbf{E}}\left[\langle C(X_k)\, Y_{k-1},\, Y_k\rangle \langle Y_k, f_r\rangle \mid \mathcal{G}_{k-1} \vee X_k\right] \\
&= \langle C(X_k)\, Y_{k-1},\, f_r\rangle\, \overline{\mathbf{E}}\left[\langle Y_k, f_r\rangle \mid \mathcal{G}_{k-1} \vee X_k\right] \\
&= \frac{1}{M} \langle C(X_k)\, Y_{k-1},\, f_r\rangle .
\end{aligned}
$$

Finally, for the first identity,

$$
\begin{aligned}
&P(Y_k = f_s \mid X_k = e_i, Y_{k-1} = f_r) \\
&= \mathbf{E}\left[\, \langle C(X_k)\, Y_{k-1},\, f_s\rangle \mid X_k = e_i, Y_{k-1} = f_r\right] \\
&= \langle C(e_i)\, f_r,\, f_s\rangle\, \mathbf{E}\left[\, 1 \mid X_k = e_i, Y_{k-1} = f_r\right] = C^i_{sr}
\end{aligned}
$$

as required. □

7.3 Filters

The 'real world' probability is P. However, our calculations will be under $\overline{P}$.

Write $Y_{0:k}$ for $\{Y_0, Y_1, \ldots, Y_k\}$ and $y_{0:k}$ for a set of observations of the Y sequence to time k. Then write $Y_{0:k} = y_{0:k}$ when $\{Y_0 = y_0, Y_1 = y_1, \ldots, Y_k = y_k\}$. Recall all the y_l are unit vectors in $\{f_1, \cdots f_M\}$.

Define

$$\alpha_i(k) = P_\theta(Y_0 = y_0, \ldots, Y_k = y_k, X_k = e_i) = \mathbf{E}_\theta\left[\prod_{l=0}^{k} \langle Y_l,\, y_l\rangle\, \langle X_k,\, e_i\rangle\right]$$

for $k = 0, 1, 2, \ldots$ and $1 \le i \le N$. Write

$$\alpha(k) = (\alpha_1(k), \cdots \alpha_N(k))^\top \in \mathbb{R}^N .$$

We also define for $k = 0, 1, 2, \ldots$ the random variables

$$q_k = \overline{\mathbf{E}}[\overline{\Lambda}_k X_k \mid \mathcal{Y}_k] = q_k(Y_{0:\,k}). \tag{7.1}$$

We now establish relationships and recursions for these quantities.

Lemma 7.3 *Given the event* $\{Y_{0:k} = y_{0:k}\}$,

$$q_k(y_{0:\,k}) = M^{k+1}\alpha(k).$$

Remark We note that, as defined in (7.1), q_k is a random variable depending on $Y_{0:k}$ while $\alpha(k)$ is a function of the observations $y_{0:k}$. In (7.1) we used q_k as a random variable and also (with an abuse of notation) as a function to express the dependence of this random variable on $Y_{0:n}$. It is in this latter sense that we use q_k is this lemma. Thus $q_k \equiv q_k(Y_{0:k})$ is a random variable while $q_k(y_{0:\,k})$ is not.

Proof For $k = 0, 1, 2, \ldots$ we have

$$\begin{aligned}\overline{\mathbf{E}}\left[\overline{\Lambda}_k X_k \mid Y_{0:k} = y_{0:k}\right] &= \overline{\mathbf{E}}\left[\overline{\mathbf{E}}\left[\overline{\Lambda}_k X_k \mid \mathcal{Y}_k\right] \mid Y_{0:k} = y_{0:k}\right] \\ &= \overline{\mathbf{E}}\left[q_k(Y_{0:k}) \mid Y_{0:k} = y_{0:k}\right] = q_k(y_{0:k})\end{aligned}$$

and as $\overline{P}(Y_{0:k} = y_{0:k}) = M^{-(k+1)}$,

$$\begin{aligned}\overline{\mathbf{E}}\left[\overline{\Lambda}_k X_k \mid Y_{0:k} = y_{0:k}\right] &= \frac{\overline{\mathbf{E}}\left[\overline{\Lambda}_k X_k\, I(Y_{0:k} = y_{0:k})\right]}{\overline{P}(Y_{0:k} = y_{0:k})} \\ &= M^{k+1}\,\overline{\mathbf{E}}\left[\overline{\Lambda}_k X_k\, I(Y_{0:k} = y_{0:k})\right] \\ &= M^{k+1}\,\mathbf{E}_\theta\left[X_k\, I(Y_{0:k} = y_{0:k})\right] = M^{k+1}\,\alpha(k)\end{aligned}$$

since, with $\alpha(k) = (\alpha_1(k), \alpha_2(k), \ldots, \alpha_N(k))^\top$,

$$\begin{aligned}\alpha_i(k) &= \langle \alpha(k), e_i\rangle \\ &= \mathbf{E}_\theta\left[\langle X_k, e_i\rangle\, I(Y_{0:k} = y_{0:k})\right] \\ &= P_\theta(Y_0 = y_0, \ldots, Y_k = y_k, X_k = e_i) .\end{aligned}$$

This proves the lemma. □

We now obtain a recurrence for the $\{\alpha(k)\}$ in the following lemma. The recurrence for $\{q_k\}$ can be derived in a similar way.

Lemma 7.4 *For $k \geq 0$, we have*

$$\alpha(k+1) = B(y_{k+1},\, y_k) \cdot A \cdot \alpha(k)$$

where

$$B(y_{k+1}, y_k) = \mathtt{diag}(B^i(y_{k+1}, y_k))$$

and

$$B^i(y_{k+1}, y_k) = \langle C(e_i)\, y_k,\, y_{k+1}\rangle = y_{k+1}^\top\, C^i\, y_k.$$

For the initial condition we have

$$\alpha(0) = \frac{\pi(0)}{M}\,.$$

Proof For $k \geq 0$ and $1 \leq i \leq N$

$$\begin{aligned}
\alpha_i(k+1) &= \mathbf{E}_\theta\left[\prod_{l=0}^{k+1}\langle Y_l, y_l\rangle\langle X_{k+1}, e_i\rangle\right] \\
&= \overline{\mathbf{E}}\left[\overline{\Lambda}_{k+1}\prod_{l=0}^{k+1}\langle Y_l, y_l\rangle\langle X_{k+1}, e_i\rangle\right] \\
&= \sum_{j=1}^{N} A_{ij}\langle C(e_i)y_k, y_{k+1}\rangle\, \alpha_j(k) \\
&= \sum_{j=1}^{N} B^i(y_{k+1}, y_k)\, A_{ij}\, \alpha_j(k)
\end{aligned}$$

establishes the recurrence. For the initial condition

$$\begin{aligned}
\alpha_i(0) &= P_\theta(Y_0 = y_0, X_0 = e_i) \\
&= \mathbf{E}_\theta[\langle Y_0, y_0\rangle\langle X_0, e_i\rangle] \\
&= \overline{\mathbf{E}}[N\langle\pi(0), X_0\rangle\langle Y_0, y_0\rangle\langle X_0, e_i\rangle] \\
&= \frac{\pi_j(0)}{M}\,.
\end{aligned}$$

Hence, the proof of Lemma 7.4 is complete. □

Corollary 7.5 *The likelihood, given $y_{0:k}$, is given by*

$$P_\theta(Y_{0:k} = y_{0:k}) = \sum_{j=1}^{N} P_\theta\left(Y_{0:k} = y_{0:k},\, X_k = e_j\right) = \sum_{j=1}^{N}\alpha_j(k)\,.$$

Corollary 7.6 *For any $k \geq 0$,*

$$q_{k+1}(y_{0:k+1}) = M \cdot B(y_{k+1}, y_k) \cdot A \cdot q_k(y_{0:k})$$

and

$$q_0 = \pi(0).$$

We shall introduce the notations for $k \leq n$

$$\gamma(k|n) = \mathbf{E}_\theta \left[X_k \,|\, \mathcal{Y}_{0:n} = y_{0:n} \right]$$

and

$$\gamma_i(k|n) = P_\theta \left(X_k = e_i \,|\, \mathcal{Y}_{0:n} = y_{0:n} \right) = \mathbf{E}_\theta \left[\langle X_k, e_i \rangle \,|\, \mathcal{Y}_{0:n} = y_{0:n} \right].$$

Corollary 7.7 *For $k \geq 0$ we have the filters*

$$\mathbf{E}_\theta \left[X_k \,|\, Y_{0:k} \right] = \frac{q_k}{\langle q_k, \mathbf{1} \rangle}.$$

This is a random variable as a function of $Y_{0:k}$. In terms of an explicit sequence of observations

$$\gamma(k|k) = \mathbf{E}_\theta \left[X_k \,|\, Y_{0:k} := y_{0:k} \right] = \frac{\alpha(k)}{\langle \alpha(k), \mathbf{1} \rangle} = \frac{q_k(y_{0:k})}{\langle q_k(y_{0:k}), \mathbf{1} \rangle}$$

as the q_k and $\alpha(k)$ are proportional to each other.

This also implies

$$\gamma_i(k|k) = P_\theta(X_k = e_i \,|\, Y_{0:k} = y_{0:k}) = \frac{\alpha_i(k)}{\langle \alpha(k), \mathbf{1} \rangle} = \frac{q_k^i(y_{0:k})}{\langle q_k(y_{0:k}), \mathbf{1} \rangle},$$

where $q_k^i = \langle q_k, e_i \rangle$.

Remark It is possible to convert the formula above for normalized filter probabilities into a nonlinear forward recursion:

$$\gamma_i(k+1|k+1) = \frac{B^i(y_{k+1}, y_k) \sum_{j=1}^N A_{ij} \, \gamma_j(k|k)}{\sum_{j,l=1}^N B^l(y_{k+1}, y_k) \, A_{lj} \, \gamma_j(k|k)}, \quad \gamma_i(0|0) = \pi(0).$$

These normalized recursions perform better in numerical calculation. This was also pointed out by Pardoux (2008).

7.4 Smoothers

We consider the process $\{\beta(k)\}$ defined by

$$\beta_i(k) = P_\theta(Y_{k+1} = y_{k+1}, \ldots, Y_n = y_n \,|\, X_k = e_i)$$

for $0 \leq k \leq n$ and each $1 \leq i \leq N$. We shall show that this process satisfies a backward recursion for which we can assign $\beta_i(n) = 1$ for all i.

We consider fixed point smoothers but then allow the terminal point to increase. Related symmetric filters can be found in Wall et al. (1981) and Bressler (1986).

Lemma 7.8 *For $0 \leq k \leq n-1$ and $1 \leq i \leq N$, let*

$$v^i_{k,n} = \overline{\mathbf{E}}\left[\overline{\Lambda}_{k+1,\,n} \,|\, \mathcal{Y}_n \vee \{X_k = e_i\}\right] = v^i_{k,n}\left(Y_{0:n}\right)$$

where we write

$$\overline{\Lambda}_{r,\,s} = \prod_{l=r}^{s} \overline{\lambda}_l\,.$$

Then

$$v^i_{k,n}\left(y_{0:n}\right) = M^{n-k}\,\beta_i(k).$$

Proof As in Lemma 7.4,

$$\begin{aligned} v^i_{k,n}\left(y_{0:n}\right) &= \overline{\mathbf{E}}\left[\overline{\Lambda}_{k+1,\,n} \,|\, Y_{0:n} = y_{0:n} \,\&\, X_k = e_i\right] \\ &= \overline{\mathbf{E}}\left[\overline{\Lambda}_{k+1,\,n} \,|\, Y_{k:n} = y_{k:n} \,\&\, X_k = e_i\right]. \end{aligned}$$

Also

$$\begin{aligned} \beta_i(k) &= P_\theta(Y_{k+1:n} = y_{k+1:n} \,|\, X_k = e_i) \\ &= \mathbf{E}_\theta\left[\prod_{l=k+1}^{n} \langle Y_l,\, y_l\rangle \,\middle|\, X_k = e_i\right] \\ &= v^i_{k,n}\left(y_{0:n}\right)\, M^{-n+k}\,\frac{\overline{\mathbf{E}}\left[\overline{\Lambda}_k\,\langle X_k, e_i\rangle\right]}{\overline{\mathbf{E}}\left[\overline{\Lambda}_k\,\langle X_k, e_i\rangle\right]} \\ &= v^i_{k,n}\left(y_{0:n}\right)\, M^{-n+k}, \end{aligned}$$

so the lemma is proved. □

Lemma 7.9 *Given a matrix $A = (A_{ij})$, for any $0 \leq k \leq n-1$,*

$$v_{k,n} = MA^\top B(Y_{k+1}, Y_k)v_{k+1,n} \tag{7.2}$$

where $v_{k,n}$ denotes $(v^1_{k,n}, v^2_{k,n}, \ldots, v^N_{k,n})^\top$ and $v_{n,n} = \mathbf{1}$, $\mathbf{1} \in \mathbb{R}^n$.

Proof We first show that

$$v_{n-1,n} = MA^\top B(Y_n, Y_{n-1})\mathbf{1} \tag{7.3}$$

which shows that in the recurrence we may take $v_{n,n} = \mathbf{1}$. From its definition,

$$\begin{aligned} v^i_{n-1,\,n} &= \overline{\mathbf{E}}\left[\Lambda_{n,n} \mid \mathcal{Y}_n \vee \{X_{n-1} = e_i\}\right] \\ &= \overline{\mathbf{E}}\left[MN\langle AX_{n-1}, X_n\rangle \langle C(X_n)Y_{n-1}, Y_n\rangle \mid \mathcal{Y}_n \vee \{X_{n-1} = e_i\}\right] \\ &= \sum_{j=1}^{N} M\, A^\top_{ij}\, B^j(Y_n, Y_{n-1})\,. \end{aligned}$$

This establishes (7.3). We now let $0 \le k \le n-2$ and proceed similarly:

$$\begin{aligned} v^i_{k,\,n} &= \overline{\mathbf{E}}\left[\Lambda_{k+1,\,n} \mid \mathcal{Y}_n \vee \{X_k = e_i\}\right] \\ &= \overline{\mathbf{E}}\left[\Lambda_{k+2,\,n}\, M\, N\, \langle AX_k,\, X_{k+1}\rangle\, \langle C(X_{k+1})\, Y_k,\, Y_{k+1}\rangle \mid \mathcal{Y}_n \vee \{X_k = e_i\}\right] \\ &= \sum_{j=1}^{N} M\, \langle Ae_i,\, e_j\rangle\, \langle C(e_j)\, Y_k,\, Y_{k+1}\rangle\, v^j_{k+1,n} \\ &= M \sum_{j=1}^{N} A^\top_{ij}\, B^j(Y_{k+1}, Y_k)\, v^j_{k+1,n} \end{aligned}$$

and the recurrence is established. □

Corollary 7.10 *For any* $0 \le k \le n-1$,

$$\beta(k) = A^\top B(y_{k+1}, y_k)\, \beta(k+1) \tag{7.4}$$

where $\beta(n) = \mathbf{1}$ *and* $\mathbf{1} \in \mathbb{R}^n$.

Proof This follows from the previous two lemmas. □

Remark A disadvantage with the backward recurrence dynamics given in (7.2) and (7.3) is that, by taking taking $\beta(n) = v_{n,n} = \mathbf{1} \in \mathbb{R}^N$, when a new observation is obtained at time $n+1$ a new β (or v) must be recalculated starting with $\beta(n+1) = v_{n+1,n+1} = \mathbf{1} \in \mathbb{R}^N$. This can be circumvented by storing the matrix

$$\Phi_{k,n} = A^\top B(y_{k+1},\, y_k)\, A^\top B(y_{k+2},\, y_{k+1}) \cdots A^\top B(y_n,\, y_{n-1}).$$

When the new observation y_{n+1} is obtained we can calculate

$$\Phi_{k,n+1} = \Phi_{k,n}\, A^\top B(y_{n+1},\, y_n)$$

and then

$$v_{k,n+1} = M\, \Phi_{k,n+1}\mathbf{1}.$$

Remark The recurrence for $\beta(k)$ has better scaling than the one for $v_{k,n}$.

We now seek formulas for smoothers. That is, we compute for $0 \le k \le n$ and $1 \le i \le N$

$$\gamma_i(k|n) = P_\theta\left(X_k = e_i \,|\, Y_{0:n} = y_{0:n}\right).$$

Note that

$$\begin{aligned} P_\theta\left(X_k = e_i \,|\, Y_{0:n} = y_{0:n}\right) &= \mathbf{E}_\theta\left[\langle X_k, e_i\rangle \,|\, Y_{0:n} = y_{0:n}\right] \\ &= \frac{\overline{\mathbf{E}}\left[\overline{\Lambda}_n \langle X_k, e_i\rangle \,|\, Y_{0:n} = y_{0:n}\right]}{\overline{\mathbf{E}}\left[\overline{\Lambda}_n \,|\, Y_{0:n} = y_{0:n}\right]} \end{aligned}$$

and

$$\begin{aligned} \overline{\mathbf{E}}\left[\overline{\Lambda}_n \langle X_k, e_i\rangle \,|\, Y_{0:n} = y_{0:n}\right] &= \overline{\mathbf{E}}\left[\overline{\mathbf{E}}\left[\overline{\Lambda}_n \langle X_k,\, e_i\rangle \,|\, \mathcal{Y}_n \vee X_k\right] \,|\, Y_{0:n} = y_{0:n}\right] \\ &= v^i_{k,n}(y_{0:n})\, \overline{\mathbf{E}}\left[\,\overline{\Lambda}_k \left\langle X_k,\, e_i\right\rangle \,|\, Y_{0:k} = y_{0:k}\right] \\ &= v^i_{k,n}(y_{0:n})\, q^i_k(y_{0:k}) \\ &= M^{n+1} \beta_i(k)\, \alpha_i(k)\,. \end{aligned}$$

Also, as

$$\overline{\mathbf{E}}\left[\overline{\Lambda}_n \,|\, Y_{0:n} = y_{0:n}\right] = \sum_{i=1}^{N} \overline{\mathbf{E}}\left[\overline{\Lambda}_n \langle X_k,\, e_i\rangle \,|\, Y_{0:n} = y_{0:n}\right],$$

the next lemma follows.

Lemma 7.11 *For $0 \le k \le n$ and $1 \le i \le N$*

$$\begin{aligned} P_\theta\left(X_k = e_i \,|\, Y_{0:n} = y_{0:n}\right) &= \frac{\beta_i(k)\, \alpha_i(k)}{\sum_{j=1}^N \beta_j(k)\, \alpha_j(k)} \\ &= \frac{v^i_{k,n}(y_{0:n})\, q^i_k(y_{0:k})}{\sum_{j=1}^N v^i_{k,n}(y_{0:n})\, q^i_k(y_{0:k})}\,. \end{aligned}$$

Lemma 7.12

$$P_\theta(Y_{0:n} = y_{0:n} \,|\, X_{0:n}) = \langle DX_0,\, y_0\rangle \prod_{l=1}^{n} \langle C(X_l)\, y_{l-1},\, y_l\rangle\,.$$

Proof The left-hand side of this equality can be computed via Bayes' Conditional Expectation Lemma (Appendix C):

$$\begin{aligned} P_\theta(Y_{0:n} = y_{0:n} \,|\, X_{0:n}) &= \mathbf{E}_\theta\left[\prod_{l=0}^{n} \langle Y_l,\, y_l\rangle \,\middle|\, X_{0:n}\right] \\ &= \frac{\overline{\mathbf{E}}\left[\overline{\Lambda}_n \prod_{l=0}^{n} \langle Y_l,\, y_l\rangle \,\middle|\, X_{0:n}\right]}{\overline{\mathbf{E}}\left[\overline{\Lambda}_n \,\middle|\, X_{0:n}\right]}. \end{aligned}$$

We now concentrate on the numerator as the expression for the denominator then follows.

$$\begin{aligned}
&\overline{\mathbf{E}}\left[\overline{\Lambda}_n \prod_{l=0}^{n}\langle Y_l,\, y_l\rangle \,\middle|\, X_{0:n}\right] \\
&= \mathbf{E}\Bigg[M^{n+1}\, N^{n+1}\,\langle \pi(0),\, X_0\rangle\,\langle DX_0,\, Y_0\rangle \prod_{l=1}^{n}\langle AX_{l-1},\, X_l\rangle \\
&\qquad \times \prod_{l=1}^{n}\langle C(X_l)Y_{l-1},\, Y_l\rangle \prod_{l=0}^{n}\langle Y_l,\, y_l\rangle \,\Bigg|\, X_{0:n}\Bigg] \\
&= M^{n+1}\, N^{n+1}\,\langle \pi(0),\, X_0\rangle\,\langle DX_0,\, y_0\rangle \\
&\qquad \times \prod_{l=1}^{n}\langle AX_{l-1},\, X_l\rangle\,\langle C(X_l)y_{l-1},\, y_l\rangle\, \overline{\mathbf{E}}\left[\prod_{l=0}^{n}\langle Y_l,\, y_l\rangle \,\middle|\, X_{0:n}\right] \\
&= N^{n+1}\,\langle \pi(0),\, X_0\rangle\,\langle DX_0,\, y_0\rangle \prod_{l=1}^{n}\langle AX_{l-1},\, X_l\rangle\,\langle C(X_l)y_{l-1},\, y_l\rangle\,.
\end{aligned}$$

We can now deduce from this that

$$\overline{\mathbf{E}}\left[\overline{\Lambda}_n \mid X_{0:n}\right] = N^{n+1}\,\langle \pi(0),\, X_0\rangle \prod_{l=1}^{n}\langle AX_{l-1},\, X_l\rangle\,.$$

This can be seen as follows. Set $y_n = f_1, f_2, \ldots, f_M$ in the above and then sum the results. Repeat this process successively for $y_{n-1}, y_{n-2}, \ldots, y_0$. The lemma now follows. □

Remark This also establishes the claim in the introduction to this chapter that, given the events $X_{0:n}$, the events $\{Y_k = y_k\}$ for $0 \le k \le n$ are no longer independent in this model. See Koski (2001), §13.3.

Lemma 7.13 (The Baum identity) *For $0 \le k \le n$ and $1 \le i \le N$ we have*

$$P_\theta\left(X_k = e_i,\, Y_{0:n} = y_{0:n}\right) = \alpha_i(k)\,\beta_i(k).$$

Remark This is a consequence of the proof of Lemma 7.12.

One consequence of Lemma 7.13 is a so-called Derin's formula (see Koski, 2001, Chapter 13). This is the backward recursion for the normalized smoother.

Lemma 7.14 (Derin's formula) *For all $1 \le j \le N$ and $0 \le k \le n-1$,*

$$\gamma_i(k|n) = \gamma_i(k|k) \sum_{j=1}^{N} a_{ji}\,\frac{\gamma_j(k+1|n)}{\gamma_j(k+1|k)}, \quad \gamma_j(n|n) = 1.$$

The relative merits of this recursion and the previous one are discussed in Devijver (1985).

Proof Let $1 \le j \le N$ and $0 \le k \le n-1$. Then

$$\begin{aligned}\gamma_i(k|n) &= \sum_{j=1}^{N} P_\theta\left(X_k = e_i,\, X_{k+1} = e_j \,|\, Y_{0:n} = y_{0:n}\right) \\ &= \sum_{j=1}^{N} \gamma_i(k+1|n) \left[\frac{P_\theta\left(X_k = e_i,\, X_{k+1} = e_j,\, Y_{0:n} = y_{0:n}\right)}{P_\theta\left(X_{k+1} = e_j,\, Y_{0:n} = y_{0:n}\right)}\right].\end{aligned}$$

But

$$\begin{aligned}&P_\theta\left(X_k = e_i,\, X_{k+1} = e_j,\, Y_{0:n} = y_{0:n}\right) \\ &\quad = P_\theta\left(Y_{0:n} = y_{0:n} \,|\, X_k = e_i,\, X_{k+1} = e_j\right) P_\theta\left(X_k = e_i,\, X_{k+1} = e_j\right) \\ &\quad = a_{ji}\, P_\theta\left(Y_{0:n} = y_{0:n} \,|\, X_k = e_i,\, X_{k+1} = e_j\right) P_\theta\left(X_k = e_i\right).\end{aligned}$$

We use Lemma 7.12 to compute $P_\theta\left(Y_{0:n} = y_{0:n} \,|\, X_k = e_i,\, X_{k+1} = e_j\right)$. In fact

$$\begin{aligned}&\mathbf{E}_\theta\left[\prod_{l=0}^{n}\langle Y_l,\, y_l\rangle \,\middle|\, X_k = e_i,\, X_{k+1} = e_j\right] \\ &\quad = \mathbf{E}_\theta\left[\mathbf{E}_\theta\left[\prod_{l=0}^{n}\langle Y_l, y_l\rangle \,\middle|\, \mathcal{F}_n\right] \,\middle|\, X_k = e_i,\, X_{k+1} = e_j\right] \\ &\quad = P_\theta\left(Y_{0:k} = y_{0:k} \,|\, X_k = e_i\right) \mathbf{E}_\theta\left[\prod_{l=k+1}^{n}\langle C(X_l) y_{l-1},\, y_l\rangle \,\middle|\, X_{k+1} = e_j\right].\end{aligned}$$

This then implies that

$$\begin{aligned}&P_\theta\left(Y_{0:n} = y_{0:n} \,|\, X_k = e_i,\, X_{k+1} = e_j\right) P_\theta\left(X_k = e_i\right) \\ &\quad = \gamma_i(k|k)\, P_\theta\left(Y_{0:k} = y_{0:k}\right) \mathbf{E}_\theta\left[\prod_{l=k+1}^{n}\langle C(X_l)\, y_{l-1},\, y_l\rangle \,\middle|\, X_{k+1} = e_j\right].\end{aligned}$$

A very similar calculation shows that

$$P_\theta\left(X_{k+1} = e_j,\, Y_{0:n} = y_{0:n}\right)$$

is the same as

$$\gamma_i(k+1|k)\, P_\theta\left(Y_{0:k} = y_{0:k}\right) \mathbf{E}_\theta\left[\prod_{l=k+1}^{n}\langle C(X_l)\, y_{l-1},\, y_l\rangle \,\middle|\, X_{k+1} = e_j\right].$$

In fact, the two expressions

$$P_\theta\left(X_{k+1}=e_j,\,Y_{0:n}=y_{0:n}\right)$$
$$=P_\theta\left(Y_{0:n}=y_{0:n}\,|\,X_{k+1}=e_j\right)\,P_\theta\left(X_{k+1}=e_j\right)$$

and

$$P_\theta\left(Y_{0:n}=y_{0:n}\,|\,X_{k+1}=e_j\right)$$

are equivalent to saying

$$P_\theta\left(Y_{0:k}=y_{0:k}\,|\,X_{k+1}=e_j\right)\mathbf{E}_\theta\left[\prod_{l=k+1}^{n}\langle C(X_l)\,y_{l-1},\,y_l\rangle\,\middle|\,X_{k+1}=e_j\right]$$

and

$$P_\theta\left(Y_{0:k}=y_{0:k}|X_{k+1}=e_j\right)P_\theta\left(X_{k+1}=e_j\right)=\gamma_j(k+1|k)P_\theta\left(Y_{0:k}=y_{0:k}\right).$$

The Derin formula then follows. □

7.5 The Viterbi algorithm

The Viterbi algorithm is used to determine the most likely sequence of the hidden states consistent with a sequence of observations. An alternative approach is to use smoothers. With this approach $i=i_k^*$ is chosen to maximize $\gamma_i(k|n)$ for each $0\le k\le n$ (if there are $n+1$ observations $y_{0:n}$). However, as Rabiner (1989) rightly points out, this smoother based 'decoding' may produce an infeasible path unless all the elements of matrix A are positive or all components of the distribution vector δ are positive. We could suggest that a tie in the Viterbi algorithm could be broken using the corresponding smoother estimates, but this is unlikely to be successful without considering further observations. This can be performed using the updating procedure discussed after Corollary 7.10, above. This updating could be continued till the tie is broken.

We can then estimate, or decode, the hidden chain in a backward recursive manner. Find i_n^* to maximize $\delta_n(i)$. Then, if we know i_{k+1}^*, take $i_k^*=\psi_{k+1}(i_{k+1}^*)$. The Viterbi estimate for the most likely path given $y_{0:n}$ is then

$$e_{i_0^*},e_{i_1^*},e_{i_2^*},\ldots,e_{i_n^*}$$

Lemma 7.15 *For $0\le k\le n$ and $1\le i\le N$, define*

$$\delta_k(i)=\max_{x_{0:k-1}\in\mathcal{S}^k}P_\theta(Y_{0:\,k}=y_{0:\,k},X_{0:\,k-1}=x_{0:\,k-1},X_k=e_i)$$

where $\mathcal{S} = \{e_1, \dots, e_N\}$.

Write $\delta_k = (\delta_k(1), \dots, \delta_k(N))^\top$. *Then we have* $\delta_0 = \pi(0)/M$ *and*

$$\begin{aligned}\delta_{k+1}(j) &= \langle C(e_j)y_k, y_{k+1}\rangle \max_{1\le i\le N}(A_{ji}\,\delta_k(i)) \\ &= B^j(y_{k+1}, y_k) \max_{1\le i\le N}(A_{ji}\,\delta_k(i)).\end{aligned}$$

Remark We can define $\psi_{k+1}(j) = i^*$ if

$$A_{ji^*}\,\delta_k(i^*) = \max_{1\le i\le N}(A_{ji}\,\delta_k(i)).$$

Proof We first derive the formula for δ_0. In fact under $\theta = (\pi(0), A, C)$,

$$\begin{aligned}\delta_0(i) &= P_\theta\left(Y_0 = y_0,\, X_0 = e_i\right) \\ &= \mathbf{E}_\theta[\langle Y_0, y_0\rangle\,\langle X_0,\, e_i\rangle] \\ &= \overline{\mathbf{E}}[\overline{\Lambda}_0\langle Y_0, y_0\rangle\,\langle X_0,\, e_i\rangle] \\ &= \overline{\mathbf{E}}[N\,\langle \pi(0),\, X_0\rangle\,\langle Y_0, y_0\rangle\,\langle X_0,\, e_i\rangle = \frac{\pi_i(0)}{M}.\end{aligned}$$

In general,

$$\begin{aligned}&\delta_{k+1}(j) \\ &= \max_{x_0,\dots,x_k} \mathbf{E}_\theta\left[\prod_{l=0}^{k}\langle X_l, x_l\rangle\,\langle X_{k+1}, e_j\rangle \prod_{l=0}^{k+1}\langle Y_l, y_l\rangle\right] \\ &= \max_{x_0,\dots,x_k} \overline{\mathbf{E}}\left[\overline{\Lambda}_{k+1}\prod_{l=0}^{k}\langle X_l, x_l\rangle\,\langle X_{k+1}, e_j\rangle \times \prod_{l=0}^{k+1}\langle Y_l, y_l\rangle\right] \\ &= \max_{x_0,\dots,x_k} \overline{\mathbf{E}}\left[\overline{\Lambda}_k MN\langle C(X_{k+1})Y_k, , Y_{k+1}\rangle\,\langle AX_k, X_{k+1}\rangle \cdot \prod_{l=0}^{k}\langle X_l, x_l\rangle\right. \\ &\qquad \left.\times\ \langle X_{k+1}, e_j\rangle \prod_{l=0}^{k+1}\langle Y_l, y_l\rangle\right] \\ &= \langle C(e_j)y_k, y_{k+1}\rangle \max_{x_0,\dots,x_k} \overline{\mathbf{E}}\left[\overline{\Lambda}_k MN\langle AX_k, e_j\rangle \prod_{l=0}^{k}\langle X_l, x_l\rangle\right. \\ &\qquad \left.\times \prod_{l=0}^{k}\langle Y_l, y_l\rangle\,\langle X_{k+1}, e_j\rangle\,\langle Y_{k+1}, y_{k+1}\rangle\right] \\ &= B^j(y_{k+1}, y_k) \max_{1\le i\le N}\left[A_{ji} \max_{x_0,\dots,x_{k-1}} \mathbf{E}_\theta\left[\langle X_k, e_i\rangle \prod_{l=0}^{k}\langle X_l, x_l\rangle\,\langle Y_l, y_l\rangle\right]\right] \\ &= B^j(y_{k+1}, y_k) \max_{1\le i\le N}\left[A_{ji}\,\delta_k(i)\right].\end{aligned}$$

This completes the proof of the lemma. □

Remark It is also possible to give an additive version of the Viterbi algorithm. In certain applications this may have better numerical properties. If we set

$$\tilde{\delta}_k(i) = \max_{x_{0:k-1} \in \mathcal{S}^k} \log P_\theta(Y_{0:k} = y_{0:k}, X_{0:k-1} = x_{0:k-1}, X_k = e_i)$$

then we can, as above, derive the recursion

$$\tilde{\delta}_{k+1}(j) = \log\langle C(e_j)\, y_k,\, y_{k+1}\rangle + \max_{1 \le i \le N} \left\{ \log A_{ji} + \tilde{\delta}_k(i) \right\}$$

and for initialization, we use

$$\tilde{\delta}_0(i) = \log \pi_i(0) - \log M$$

for which we would need $\pi_i(0) > 0$ for all i.

7.6 Parameter estimation by the EM algorithm

This is similar to procedures in Chapter 6. Let $\theta = (\pi(0), D, A, C)$ where $C = C^i_{rs}$ is an initial estimation of the parameters. Some initial, randomly chosen, values for θ can be chosen.

We seek $\theta = (\pi(0), D, A, C)$ so that, given observations

$$y_{0:n} = (y_0, y_1, \ldots, y_n),$$

the probability

$$P_\theta(Y_{0:n} = y_{0:n})$$

is a maximum. In other words, we seek a maximum likelihood estimator for θ.

This problem is rarely solvable exactly. The EM algorithm provides a sequence $\{\hat{\theta}_p\}$ of parameters that provide increasing values for the likelihood.

Lemma 7.16 *Given* $\theta = (\pi(0), D, A, C)$,

$$P_\theta(Y_{0:n} = y_{0:n}) = \frac{1}{M^{n+1}} \overline{\mathbf{E}}\left[\overline{\Lambda}_n(\theta) \,\middle|\, Y_{0:n} = y_{0:n} \right],$$

where

$$\overline{\Lambda}_n(\theta) =$$

$$M^{n+1} N^{n+1} \langle \pi(0), X_0 \rangle \langle D X_0, Y_0 \rangle \prod_{l=1}^{n} \langle A X_{l-1}, X_l \rangle \prod_{l=1}^{n} \langle C(X_l) Y_{l-1}, Y_l \rangle.$$

Proof We have

$$P_\theta\left(Y_{0:n} = y_{0:n}\right) = \mathbf{E}_\theta\left[\prod_{l=0}^{n}\langle Y_l, y_l\rangle\right]$$
$$= \overline{\mathbf{E}}\left[\overline{\Lambda}_n(\theta)\prod_{l=0}^{n}\langle Y_l, y_l\rangle\right]$$
$$= \frac{1}{M^{n+1}}\,\overline{\mathbf{E}}\left[\overline{\Lambda}_n(\theta)\,\middle|\, Y_{0:n} = y_{0:n}\right]$$

and we are done. □

We recall Lemma 6.2:

Lemma 7.17 *Let*

$$Q(\theta', \theta) \equiv \mathbf{E}_\theta\left[\log\frac{\overline{\Lambda}_n(\theta')}{\overline{\Lambda}_n(\theta)}\,\middle|\, Y_{0:n} = y_{0:n}\right],$$

then

$$\log L_{\theta'} - \log L_\theta \geq Q(\theta', \theta),$$

where $L_\theta := P_\theta\left(Y_{0:n} = y_{0:n}\right)$ *is the likelihood with parameter* θ.

The function Q is called the *pseudo-log-likelihood.* The prime simply indicates a different choice of parameters: thus, $\theta' = (\pi'(0), D', A', C')$.

7.7 Steps of the EM algorithm

Step 1: Choose $\theta = \hat{\theta}_0$.
Make a random choice for the initial value of the parameters.

Step 2 (The E step): Given $\theta = \hat{\theta}_p$, compute $Q(\theta', \theta)$.

Step 3 (The M step): Maximize $Q(\theta', \theta)$ with respect to θ' and get $\hat{\theta}_{p+1}$.

We note that $\theta' = \theta$ implies $Q(\theta', \theta) = 0$ and so $Q(\hat{\theta}_{p+1}, \theta) \geq 0$. Usually, $Q(\theta', \theta) > 0$ with $\theta' = \hat{\theta}_{p+1}$ and this will imply $L_{\theta'} > L_\theta$.

Step 4: Replace $\hat{\theta}_p$ with $\hat{\theta}_{p+1}$ and return to Step 2.

The E step

We need only maximize

$$\mathbf{E}_\theta\left[\log\overline{\Lambda}_n(\theta')\,\middle|\,Y_{0:n}=y_{0:n}\right]$$

because

$$Q(\theta',\theta)=\mathbf{E}_\theta\left[\log\overline{\Lambda}_n(\theta')\,\middle|\,Y_{0:n}=y_{0:n}\right]-\mathbf{E}_\theta\left[\log\overline{\Lambda}_n(\theta)\,\middle|\,Y_{0:n}=y_{0:n}\right].$$

Also, note that

$$\overline{\Lambda}_n(\theta')=M^{n+1}N^{n+1}\cdot\overline{\Lambda}_n^{\pi}(\theta')\cdot\overline{\Lambda}_n^{D}(\theta')\cdot\overline{\Lambda}_n^{A}(\theta')\cdot\overline{\Lambda}_n^{C}(\theta'),$$

where

$$\begin{aligned}
\overline{\Lambda}_n^{\pi}(\theta')&=\langle\pi'(0),X_0\rangle,\\
\overline{\Lambda}_n^{D}(\theta')&=\langle D'X_0,Y_0\rangle,\\
\overline{\Lambda}_n^{A}(\theta')&=\prod_{l=1}^{n}\langle A'X_{l-1},X_l\rangle,\\
\overline{\Lambda}_n^{C}(\theta')&=\prod_{l=1}^{n}\langle C'(X_l)Y_{l-1},Y_l\rangle.
\end{aligned}$$

Then

$$\begin{aligned}
&\mathbf{E}_\theta\left[\log\overline{\Lambda}_n(\theta')\,\middle|\,Y_{0:n}=y_{0:n}\right]\\
&\quad=\text{const.}+\mathbf{E}_\theta\left[\log\overline{\Lambda}_n^{\pi}(\theta')\,\middle|\,Y_{0:n}=y_{0:n}\right]+\mathbf{E}_\theta\left[\log\overline{\Lambda}_n^{D}(\theta')\,\middle|\,Y_{0:n}=y_{0:n}\right]\\
&\qquad+\mathbf{E}_\theta\left[\log\overline{\Lambda}_n^{A}(\theta')\,\middle|\,Y_{0:n}=y_{0:n}\right]+\mathbf{E}_\theta\left[\log\overline{\Lambda}_n^{C}(\theta')\,\middle|\,Y_{0:n}=y_{0:n}\right].
\end{aligned}$$

Given a sequence of observations $y_{0:n}$ we now compute each of the four terms on the right-hand side of this expression. Set

$$\begin{aligned}
T_1(\theta')&=\mathbf{E}_\theta\left[\log\langle\pi'(0),X_0\rangle\,\middle|\,Y_{0:n}=y_{0:n}\right]\\
T_2(\theta')&=\mathbf{E}_\theta\left[\log\langle D'X_0,Y_0\rangle\,\middle|\,Y_{0:n}=y_{0:n}\right]\\
T_3(\theta')&=\sum_{l=1}^{n}\mathbf{E}_\theta\left[\log\langle A'X_{l-1},X_l\rangle\,\middle|\,Y_{0:n}=y_{0:n}\right]\\
T_4(\theta')&=\sum_{l=1}^{n}\mathbf{E}_\theta\left[\log\langle C'(X_l)Y_{l-1},Y_l\rangle\,\middle|\,Y_{0:n}=y_{0:n}\right].
\end{aligned}$$

Formula for $T_1(\theta')$. We have

$$T_1(\theta') = \mathbf{E}_\theta \left[\sum_{i=1}^N \langle X_0, e_i \rangle \log \langle \pi'(0), X_0 \rangle \,\middle|\, Y_{0:n} = y_{0:n} \right]$$
$$= \sum_{i=1}^N \log \pi'_i(0)\, \mathbf{E}_\theta \left[\langle X_0, e_i \rangle \,|\, Y_{0:n} = y_{0:n} \right]$$
$$= \sum_{i=1}^N \gamma_i(0) \log \pi'_i(0)\,,$$

where

$$\gamma_i(0) = \frac{\alpha_i(0)\beta_i(0)}{\sum_{i=1}^N \alpha_i(0)\beta_i(0)} = \frac{\alpha_i(0)\beta_i(0)}{L_\theta},$$

using smoothers.

Formula for $T_2(\theta')$. We also have

$$T_2(\theta') = \mathbf{E}_\theta \left[\sum_{i=1}^N \sum_{r=1}^M \langle X_0, e_i \rangle \langle Y_0, f_r \rangle \log \langle D' X_0, Y_0 \rangle \,\middle|\, Y_{0:n} = y_{0:n} \right]$$
$$= \sum_{i=1}^N \sum_{r=1}^M \log D'_{ri}\, \mathbf{E}_\theta \left[\langle X_0, e_i \rangle \langle Y_0, f_r \rangle \,|\, Y_{0:n} = y_{0:n} \right]$$
$$= \sum_{i=1}^N \sum_{r=1}^M \gamma_i(0) \langle y_0, f_r \rangle \log D'_{ri}\,.$$

Formula for $T_3(\theta')$. Similarly,

$$T_3(\theta') = \sum_{l=1}^n \sum_{i=1}^N \sum_{j=1}^N \log A'_{ji}\, \mathbf{E}_\theta \left[\langle X_l, e_j \rangle \langle X_{l-1}, e_i \rangle \,|\, Y_{0:n} = y_{0:n} \right]$$
$$= \sum_{l=1}^n \sum_{i=1}^N \sum_{j=1}^N \xi_{l-1}(j,i) \log A'_{ji}\,.$$

We need the following result to compute $T_3(\theta')$.

Lemma 7.18 *Let $0 \le k \le n-1$, $1 \le i, j \le N$. Then*

$$\xi_k(j,i) = P_\theta \left(X_{k+1} = e_j,\, X_k = e_i,\, |\, Y_{0:n} = y_{0:n} \right)$$

is given by

$$\xi_k(j,i) = A_{j\,i}\, P_\theta \left(X_k = e_i \,|\, Y_{0:n} = y_{0:n} \right) = A_{j\,i}\, \gamma_i(k|n)\,.$$

Proof We use

$$\xi_k(j,i) = P_\theta\left(X_{k+1} = e_j | X_k = e_i, Y_{0:n} = y_{0:n}\right) P_\theta\left(X_k = e_i, | Y_{0:n} = y_{0:n}\right)$$

and

$$\begin{aligned} P_\theta\big(X_{k+1} = e_j \,|\, X_k = e_i, Y_{0:n} = y_{0:n}\big) &\\ &= \mathbf{E}\left[\langle X_{k+1},\, e_j\rangle \,|\, X_k = e_i, Y_{0:n} = y_{0:n}\right] \\ &= \mathbf{E}\left[\langle A\, X_k + M_{k+1},\, e_j\rangle \,|\, X_k = e_i, Y_{0:n} = y_{0:n}\right] \\ &= A_{j\,i} + \langle \mathbf{E}\left[M_{k+1} \,|\, X_k = e_i, Y_{0:n} = y_{0:n}\right],\, e_j\rangle = A_{j\,i} \end{aligned}$$

and we are done. □

Formula for $T_4(\theta')$. Furthermore,

$$\begin{aligned} &T_4(\theta') \\ &= \sum_{l=1}^{n} \mathbf{E}_\theta[\log\langle C(X_l)Y_{l-1}, Y_l\rangle \,|\, Y_{0:\,n} = y_{0:\,n}] \\ &= \sum_{l=1}^{N}\sum_{j=1}^{N}\sum_{r=1}^{M}\sum_{s=1}^{M} \mathbf{E}_\theta[\langle X_l, e_j\rangle\langle Y_{l-1}, f_r\rangle\langle Y_l, f_s\rangle \times \\ &\qquad \times \log\langle C(X_l)Y_{l-1}, Y_l\rangle \,|\, Y_{0:\,n} = y_{0:\,n}] \\ &= \sum_{l=1}^{N}\sum_{j=1}^{N}\sum_{r=1}^{M}\sum_{s=1}^{M} \log(C^j_{sr})\mathbf{E}_\theta[\langle X_l, e_j\rangle \,|\, Y_{0:\,n} = y_{0:\,n}]\langle y_{l-1}, f_r\rangle\langle y_l, f_s\rangle \\ &= \sum_{j=1}^{N}\sum_{r=1}^{M}\left[\sum_{s=1}^{M}\sum_{l=1}^{N} \gamma_j(l)\langle y_{l-1}, f_r\rangle\langle y_l, f_s\rangle \log C^j_{sr}\right]. \end{aligned}$$

The **E step** is completed.

The M step

We proceed to the **M step**, cf. Lemma 6.3.

Lemma 7.19 *Let* $a_1, a_2, \ldots, a_n \geq 0$ *with*

$$A = \sum_{i=1}^{n} a_i > 0.$$

Let

$$\mathcal{U} = \left\{ \mathbf{x} \in \mathbb{R}^n \,\middle|\, x_i > 0 \text{ for all } i \text{ with } \sum_{i=1}^{n} x_i = 1 \right\}.$$

Then the supremum of

$$f(\mathbf{x}) = \sum_{i=1}^{n} a_i \log x_i, \quad \mathbf{x} \in \mathcal{U}$$

is attained at $\mathbf{x} = \hat{\mathbf{x}}$, *where*

$$\hat{x}_i = \begin{cases} 0 & \text{if } a_i = 0 \\ a_i/A & \text{if } a_i > 0 \end{cases}.$$

As in Chapter 6 this enables us to write down the following maxima.

Maximization of T_1. Recall that

$$T_1(\theta') = \sum_{i=1}^{N} \gamma_i(0|n) \log \pi'_i(0).$$

The maximum is given by

$$\pi'_i(0) = \begin{cases} \gamma_i(0|n) & \text{if } \gamma_i(0|n) > 0 \\ 0 & \text{otherwise} \end{cases}$$

as

$$\sum_{i=1}^{N} \gamma_i(0|n) = \sum_{i=1}^{N} \mathbf{E}_\theta \left[\langle X_0, e_i \rangle \,\middle|\, Y_{0:n} = y_{0:n} \right] = 1.$$

So $\hat{\pi}_i(0) = \gamma_i(0|n)$ for $i = 1, \ldots, N$, and the update is given by

$$\hat{\pi}(0) = \gamma(0|n).$$

Maximization for T_2. Recall that

$$T_2(\theta') = \sum_{i=1}^{N} \sum_{r=1}^{M} \gamma_i(0|n) \langle y_0, f_r \rangle \log D'_{ri};$$

then for such i, the update for D_{ri} is given by

$$\widehat{D}_{ri} = \gamma_i(0|n) < y_0, f_r >$$

which is rather degenerate. However if we had observations for L sequences (for example in genomics examples), then the update would be

$$\hat{D}_{ri} = \frac{\sum_{m=1}^{L} \gamma_i^m(0|n) \langle y_0^m, f_r \rangle}{\sum_{m=1}^{L} \gamma_i^m(0|n)}$$

where the symbols with superscript m are values computed for sequence m of the observations for each $1 \leq m \leq L$. The other parameter estimations could similarly be modified for multiple sequence observations.

Maximization for T_3. Recall that

$$T_3(\theta') = \sum_{l=1}^{n}\sum_{i=1}^{N}\sum_{j=1}^{N} \xi_{l-1}(j,i) \log A'_{ji}$$

$$= \sum_{i=1}^{N}\left\{\sum_{j=1}^{N}\left(\sum_{l=1}^{n} \xi_{l-1}(j,i)\right) \log A'_{ji}\right\}.$$

If we assume that

$$\sum_{j=1}^{N}\sum_{l=1}^{n} \xi_{l-1}(j,i) = \sum_{l=1}^{n} \gamma_{l-1}(i|n) = \sum_{l=0}^{n-1} \gamma_l(i|n) > 0$$

then for such i, the update for A_{ji} is given by

$$\hat{A}_{ji} = \frac{\sum_{l=1}^{n} \xi_{l-1}(j,i)}{\sum_{j=1}^{N}\sum_{l=1}^{n} \xi_{l-1}(j,i)} = \frac{\sum_{l=0}^{n-1} \xi_l(j,i)}{\sum_{l=0}^{n-1} \gamma_l(i|n)}.$$

However if

$$\sum_{l=0}^{n-1} \gamma_l(i|n) = 0$$

we have no new information to update A_{ji} and so we retain the current value of A_{ji} for that value of i.

Maximization for T_4. Recall that

$$T_4(\theta') = \sum_{j=1}^{N}\sum_{r=1}^{M}\left[\sum_{s=1}^{M}\sum_{l=1}^{N} \gamma_j(l|n)\langle y_{l-1}, f_r\rangle\langle y_l, f_s\rangle \log C^j_{sr}\right].$$

For each fixed j, r, we optimize over the C^l_{sr} for $1 \leq s \leq M$ giving new estimates

$$\hat{C}^j_{sr} = \frac{\sum_{l=1}^{n} \gamma_j(l|n)\langle y_{l-1}, f_r\rangle\langle y_l, f_s\rangle}{\sum_{l=1}^{n} \gamma_j(l|n)\langle y_{l-1}, f_r\rangle}.$$

If the denominator in any of the estimators is zero, then we do not update the corresponding parameter.

Extensions

It is possible to introduce higher order models for X and Y. By a clever relabelling these can be reduced to first-order chains, though the transition matrices will have many zero entries. In particular, Y could be a chain of order $k \geq 2$. All the above results can be applied, though the notation is more cumbersome.

7.8 Exercises

Exercise 7.1 Consider a model with

$$X_{k+1} = AX_k + V_{k+1}$$
$$Y_k = C(X_{k-1})Y_{k-1} + W_k.$$

Derive the result analogous to Corollary 7.6.

Exercise 7.2 For this model derive the analog of Lemma 7.9.

8
Semi-Markov Models

8.1 Introduction

We now progress to semi-Markov chains. For a Markov chain the length of time the chain spends in any state is a geometrically distributed random variable. This property will be relaxed in this chapter leading to a definition and discussion of semi-Markov models. Various properties and parametrizations of semi-Markov models will first be described.

8.2 Semi-Markov models

We now replace the Markov chain by a semi-Markov chain. With the former, we had

$$P(X_{t+1} = e_i, \ldots, X_{t+m-1} = e_i, X_{t+m} \neq e_i \,|\, X_t = e_i) = (1 - a_{ii})\, a_{ii}^{m-1}.$$

So the duration of X_t in e_i given $X_t = e_i$ has a geometric distribution and depends only on i. We shall generalize this property.

We now define a **semi-Markov chain** (Barbu and Limnios, 2008; Çinlar, 1975; Harlamov, 2008; Howard, 1971; Jelinek and Mercer, 1980 and Koski, 2001). Let

$$T_0 < T_1 < T_2 < \cdots$$

be a point process with $T_0 = 0$ (for example the jump times of a Poisson Process). A **marked point process** is a sequence

$$\{(T_n, Z_n) \,:\, n = 0, 1, 2, \ldots\},$$

where Z_n is called a **mark**, associated with jump time (e.g. $Z_n \equiv 1$ for Poisson). Also, Z_n takes values in a **mark space**. Çinlar (1975) uses a countable set for the state space. We again take the state space to be the set of unit vectors

$$S = \{e_1, \ldots, e_N\} \subset \mathbb{R}^N.$$

Also we require

$$\begin{aligned} &P(Z_{n+1} = e_j, T_{n+1} - T_n \le t \,|\, Z_n = e_i, Z_0, \ldots, Z_{n-1}, T_0, \ldots, T_n) \\ &= P(Z_{n+1} = e_j, T_{n+1} - T_n \le t \,|\, Z_n = e_i) \equiv Q(e_j,\, e_i,\, t) \end{aligned}$$

to be independent of n. A **semi-Markov process** is a process $\{X_t \colon t \ge 0\}$ with

$$X_t = Z_n \quad \text{if} \quad T_n \le t < T_{n+1}\,.$$

Standard references are Çinlar (1975) and Howard (1971).

In our study, we shall consider discrete time with

$$T_n \in \{0, 1, 2, 3, \ldots\},\ T_{n+1} - T_n \ge 1 \text{ for all } n, \quad \text{and} \quad T_0 = 0.$$

We set

$$P(Z_{n+1} = e_j, T_{n+1} - T_n = m \,|\, Z_n = e_i) \equiv q(e_j,\, e_i,\, m).$$

and define

$$\begin{aligned} \tau_{ji}(m) &\equiv P(T_{n+1} - T_n = m \,|\, Z_{n+1} = e_j, Z_n = e_i) \\ p_{ji} &\equiv P(Z_{n+1} = e_j \,|\, Z_n = e_i) \quad \text{for } j \neq i. \end{aligned}$$

We shall assume that $p_{ii} = 0$, following Barbu and Limnios (2008), though Howard (1971) and Çinlar (1975) relax this and allow 'virtual' transitions, as they call them. We shall show that for a semi-Markov chain, we can assume without loss of generality that we only have proper transitions and $p_{ii} = 0$.

The semi-Markov chain for a given state space, can be specified in terms of $\{\tau_{ji}(m)\}$ and $\{p_{ji}\}$.

We have the following identities

$$p_{ji} = \sum_{m=1}^{\infty} \mathrm{P}(Z_{n+1} = e_j, T_{n+1} - T_n = m | Z_n = e_i) = \sum_{m=1}^{\infty} q(e_j, e_i, m) < \infty,$$

which does not depend on n.

We also have

$$\begin{aligned} &q(e_j, e_i, m) \\ &= P(Z_{n+1} = e_j, T_{n+1} - T_n = m \,|\, Z_n = e_i) \\ &= \mathbf{E}[\mathrm{I}(Z_{n+1} = e_j)\, \mathrm{I}(T_{n+1} - T_n = m) \,|\, Z_n = e_i] \end{aligned}$$

$$
\begin{aligned}
&= \mathbf{E}[\mathbf{E}[\mathrm{I}(Z_{n+1} = e_j)\,\mathrm{I}(T_{n+1} - T_n = m) \,|\, Z_{n+1}, Z_n] \,|\, Z_n = e_i] \\
&= \mathbf{E}[\mathrm{I}(Z_{n+1} = e_j)\,\mathbf{E}[\,\mathrm{I}(T_{n+1} - T_n = m) \,|\, Z_{n+1}, Z_n] \,|\, Z_n = e_i] \\
&= \mathbf{E}[\mathrm{I}(Z_{n+1} = e_j)\,\mathbf{E}[\mathrm{I}(T_{n+1} - T_n = m) \,|\, Z_{n+1} = e_j, Z_n = e_i] \,|\, Z_n = e_i] \\
&= \mathbf{E}[\mathrm{I}(T_{n+1} - T_n = m) \,|\, Z_{n+1} = e_j, Z_n = e_i]\mathbf{E}[\mathrm{I}(Z_{n+1} = e_j) \,|\, Z_n = e_i] \\
&= P(T_{n+1} - T_n = m \,|\, Z_{n+1} = e_j, Z_n = e_i)\; P(Z_{n+1} = e_j \,|\, Z_n = e_i) \\
&= \tau_{ji}(m)\, p_{ji}\,.
\end{aligned}
$$

Suppose $X_t = e_i$. We introduce the following function:

$$h_t(X_t) = k$$

when

$$X_{t-k} \neq e_i, X_{t-k+1} = e_i, \ldots, X_t = e_i.$$

It follows from its definition that

$$h_t(X_t) = 1 + \langle X_t, X_{t-1} \rangle\, h_{t-1}(X_{t-1}).$$

Example 8.1 We can check

$$h_1(X_1) = 1 + \langle X_1, X_0 \rangle.$$

If $X_1 = e_i$, either $X_0 \neq e_i$ or $X_0 = e_i$. Both sides are equal to 1 or 2.

We can also check that

$$h_2(X_2) = 1 + \langle X_2, X_1 \rangle + \langle X_2, X_1 \rangle \langle X_1, X_0 \rangle = 1 + \langle X_2, X_1 \rangle\, h_1(X_1).$$

For $X_2 = e_i$, $X_1 = e_i$, and $X_0 \neq e_i$, we get

$$1 + \langle X_2, X_1 \rangle\, h_1(X_1) = 1 + 1 = 2,$$

and for $X_2 = e_i$, $X_1 = e_i$, and $X_0 = e_i$,

$$1 + \langle X_2, X_1 \rangle\, h_1(X_1) = 3.$$

8.3 Transition probabilities for semi-Markov chains

The next result is basic for semi-Markov chains.

Theorem 8.2 *For all $1 \leq j, i \leq N$, we have*

$$
\begin{aligned}
&P\left(X_{t+1} = e_j \,|\, h_t(X_t) = k \,\vee\, X_t = e_i \,\vee\, \mathcal{F}_{t-1}^X\right) \\
&\quad = P\left(X_{t+1} = e_j \,|\, h_t(X_t) = k \,\vee\, X_t = e_i\,\right)
\end{aligned}
$$

and is not dependent on t.

Proof Because we have

$$\sum_{j=1}^{N} P\left(X_{t+1} = e_j \,|\, h_t(X_t) = k \vee X_t = e_i \vee \mathcal{F}_{t-1}^X\right) = 1$$

and

$$\sum_{j=1}^{N} P\left(X_{t+1} = e_j \,|\, h_t(X_t) = k \vee X_t = e_i\right) = 1$$

we need only prove the identity when $j \neq i$. We can write

$$\begin{aligned}
&P\big(X_{t+1} = e_j \,|\, h_t(X_t) = k \vee X_t = e_i \vee \mathcal{F}_{t-1}^X\big) \\
&= \sum_{n=0}^{\infty} P(X_{t+1} = e_j,\, T_n \leq t < T_{n+1} \,|\, \mathcal{F}_{t-1}^X,\, X_t = e_i,\, h_t(X_t) = k) \\
&= \sum_{n=0}^{\infty} P(Z_{t+1} = e_j,\, T_n \leq t < T_{n+1} \,|\, Z_{0:n-1},\, Z_n = e_i,\, T_{0:n}, \\
&\qquad\qquad T_{n+1} - T_n = k) \\
&= \sum_{n=0}^{\infty} P(Z_{t+1} = e_j,\, T_n \leq t < T_n + k \,|\, Z_{0:n-1},\, Z_n = e_i,\, T_{0:n}, \\
&\qquad\qquad T_{n+1} - T_n = k) \\
&= \sum_{n=0}^{\infty} I(T_n \leq t < T_n + k)\, P(Z_{n+1} = e_j \,|\, Z_{0:n-1}, Z_n = e_i,\, T_{0:n}, \\
&\qquad\qquad T_{n+1} - T_n = k) \\
&= \sum_{n=0}^{\infty} I(T_n \leq t < T_n + k) \times \\
&\qquad \times \frac{P(Z_{n+1} = e_j,\, T_{n+1} - T_n = k \,|\, Z_{0:n-1},\, Z_n = e_i,\, T_{0:n})}{P(T_{n+1} - T_n = k \,|\, Z_{0:n-1},\, Z_n = e_i,\, T_{0:n})}
\end{aligned}$$

and because by property of renewal process, this is

$$\begin{aligned}
&= \sum_{n=0}^{\infty} I(T_n \leq t < T_n + k)\, \frac{P(Z_{n+1} = e_j, T_{n+1} - T_n = k \,|\, Z_n = e_i)}{P(T_{n+1} - T_n = k \,|\, Z_n = e_i)} \\
&= \sum_{n=0}^{\infty} I(T_n \leq t < T_n + k)\, \frac{P(Z_{n+1} = e_j, T_{n+1} - T_n = k \,|\, T_n,\, Z_n = e_i)}{P(T_{n+1} - T_n = k \,|\, T_n,\, Z_n = e_i)} \\
&= \sum_{n=0}^{\infty} I(T_n \leq t < T_n + k)\, P(Z_{n+1} = e_j \,|\, T_n,\; Z_n = e_i, T_{n+1} - T_n = k)
\end{aligned}$$

$$= \sum_{n=0}^{\infty} P(Z_{n+1} = e_j, T_n \le t < T_n + k \,|\, T_n,\, Z_n = e_i, T_{n+1} - T_n = k)$$
$$= \sum_{n=0}^{\infty} P(Z_{n+1} = e_j, T_n \le t < T_{n+1} \,|\, T_n,\, Z_n = e_i, T_{n+1} - T_n = k)$$
$$= \sum_{n=0}^{\infty} P(X_{t+1} = e_j, T_n \le t < T_{n+1} \,|\, X_t = e_i, h_t(X_t) = k)$$
$$= P(X_{t+1} = e_j \,|\, X_t = e_i, h_t(X_t) = k)\,.$$

We now discuss some results that we have used. In the last few lines we used the observation which holds for $T_n \le t < T_{n+1}$, then conditioning on $\{T_n,\, Z_n = e_i, T_{n+1} - T_n = k\}$ is the same as conditioning on $\{X_t = e_i, h_t(X_t) = k\}$ for $X_t = e_i$ implies $Z_n = e_i$, $T_n = t - h_t(X_t) + 1$ and $T_{n+1} - T_n = k$. Conversely, $X_t = Z_n = e_i$, $h_t(X_t) = t - T_n + 1$.

We also have

$$\begin{aligned}
&P(Z_{n+1} = e_j,\, T_{n+1} - T_n = k \,|\, T_n,\, Z_n = e_i)\\
&= \mathbf{E}[\,\mathrm{I}(Z_{n+1} = e_j)\,\mathrm{I}(T_{n+1} - T_n = k) \,|\, T_n,\, Z_n = e_i]\\
&= \mathbf{E}[\,\mathbf{E}[\,\mathrm{I}(Z_{n+1} = e_j)\,\mathrm{I}(T_{n+1} - T_n = k) \,|\, Z_n,\, T_{n+1} - T_n] \,|\, T_n,\, Z_n = e_i]\\
&= \mathbf{E}[\mathrm{I}(T_{n+1} - T_n = k)\mathbf{E}[\mathrm{I}(Z_{n+1} = e_j)|Z_n,\, T_{n+1} - T_n]|T_n,\, Z_n = e_i]\\
&= \mathbf{E}[\mathrm{I}(T_{n+1} - T_n = k)\mathbf{E}[\mathrm{I}(Z_{n+1} = e_j)|Z_n = e_i, T_{n+1} - T_n = k]|T_n,\\
&\qquad\qquad Z_n = e_i]\\
&= \mathbf{E}[\mathrm{I}(Z_{n+1} = e_j)|T_n, Z_n = e_i, T_{n+1} - T_n = k]\mathbf{E}[\mathrm{I}(T_{n+1} - T_n = k)|T_n,\\
&\qquad\qquad Z_n = e_i]\\
&= P(Z_{n+1} = e_j|T_n, Z_n = e_i, T_{n+1} - T_n = k)P(T_{n+1} - T_n = k|T_n,\\
&\qquad\qquad Z_n = e_i)
\end{aligned}$$

and if $P(T_{n+1} - T_n = k \,|\, T_n,\, Z_n = e_i) > 0$, then

$$\begin{aligned}
&P(Z_{n+1} = e_j \,|\, T_n,\, Z_n = e_i,\, T_{n+1} - T_n = k)\\
&\quad = \frac{P(Z_{n+1} = e_j,\, T_{n+1} - T_n = k \,|\, T_n,\, Z_n = e_i)}{P(T_{n+1} - T_n = k \,|\, T_n,\, Z_n = e_i)}.
\end{aligned}$$

A similar calculation gives

$$\begin{aligned}
&P(Z_{n+1} = e_j \,|\; Z_n = e_i, T_{n+1} - T_n = k)\\
&\quad = \frac{P(Z_{n+1} = e_j, T_{n+1} - T_n = k \,|\, Z_n = e_i)}{P(T_{n+1} - T_n = k \,|\, Z_n = e_i)}
\end{aligned}$$

and

$$P(Z_{t+1} = e_j, \mid Z_{0:n-1}, Z_n = e_i, T_{0:n}, T_{n+1} - T_n = k)$$
$$\frac{P(Z_{n+1} = e_j, T_{n+1} - T_n = k \mid Z_{0:n-1}, Z_n = e_i, T_{0:n})}{P(T_{n+1} - T_n = k \mid Z_{0:n-1}, Z_n = e_i, T_{0:n})}.$$

We also use results like

$$\begin{aligned} &P(Z_{n+1} = e_j, T_{n+1} - T_n = k \mid Z_n = e_i) \\ &\quad = P(Z_{n+1} = e_j, T_{n+1} - T_n = k \mid T_n, Z_n = e_i). \end{aligned}$$

When $P(T_{n+1} - T_n = k \mid Z_n = e_i) = 0$, then a direct argument (see Appendix D) shows that

$$P(Z_{n+1} = e_j \mid Z_{0:n-1}, Z_n = e_i, T_{0:n}, T_{n+1} - T_n = k)$$

is the same as

$$P(Z_{n+1} = e_j \mid T_n, Z_n = e_i, T_{n+1} - T_n = k)$$

and modulo the results in the appendix, the theorem is proved. □

Notation For $1 \leq i \leq N$ and $m \geq 1$, set

$$p_i(k) = P(T_{n+1} - T_n = k \mid Z_n = e_i)$$

which is a quantity that does not depend on n. In fact

$$p_i(k) = \sum_{j=1}^{N} P(Z_{n+1} = e_j, T_{n+1} - T_n = k \mid Z_n = e_i) = \sum_{j=1}^{N} q(e_j, e_i, k)$$

where the sum is taken for $j \neq i$.

Lemma 8.3 *If $X_t = e_i$ and $P(h_t(X_t) = k | X_t = e_i) > 0$, then*

$$P(X_{t+1} \neq e_i \mid X_t = e_i, h_t(X_t) = k) = \frac{p_i(k)}{F_i(k)}$$

where

$$F_i(k) = \sum_{l=k}^{\infty} p_i(l).$$

Proof We have

$$\begin{aligned} &P(X_{t+1} \neq e_i \mid X_t = e_i, h_t(X_t) = k) \\ &\quad = \frac{P(X_{t+1} \neq e_i, h_t(X_t) = k \mid X_t = e_i)}{P(h_t(X_t) = k \mid X_t = e_i)}. \end{aligned}$$

This is because

$$
\begin{aligned}
&P(X_{t+1} \neq e_i,\, h_t(X_t) = k \,|\, X_t = e_i) \\
&\quad = \mathbf{E}[\, \mathrm{I}(X_{t+1} \neq e_i)\, \mathrm{I}(h_t(X_t) = k) \,|\, X_t = e_i] \\
&\quad = \mathbf{E}[\, \mathbf{E}[\, \mathrm{I}(X_{t+1} \neq e_i)\, \mathrm{I}(h_t(X_t) = k) \,|\, X_t,\, h_t(X_t)] \,|\, X_t = e_i] \\
&\quad = \mathbf{E}[\, \mathrm{I}(h_t(X_t) = k)\, \mathbf{E}[\, \mathrm{I}(X_{t+1} \neq e_i) \,|\, X_t,\, h_t(X_t)] \,|\, X_t = e_i] \\
&\quad = \mathbf{E}[\, \mathrm{I}(h_t(X_t) = k)\, \mathbf{E}[\, \mathrm{I}(X_{t+1} \neq e_i) \,|\, X_t = e_i,\, h_t(X_t) = k\,] \,|\, X_t = e_i] \\
&\quad = \mathbf{E}[\, \mathrm{I}(X_{t+1} \neq e_i) \,|\, X_t = e_i,\, h_t(X_t) = k\,]\, \mathbf{E}[\, \mathrm{I}(h_t(X_t) = k) \,|\, X_t = e_i] \\
&\quad = P(X_{t+1} \neq e_i \,|\, X_t = e_i,\, h_t(X_t) = k)\, P(h_t(X_t) = k \,|\, X_t = e_i).
\end{aligned}
$$

Calculating the numerator,

$$
\begin{aligned}
&P(X_{t+1} \neq e_i,\, h_t(X_t) = k \,|\, X_t = e_i) \\
&\quad = \sum_{n=0}^{\infty} P(X_{t+1} \neq e_i,\, h_t(X_t) = k,\, T_n \leq t < T_{n+1} \,|\, X_t = e_i) \\
&\quad = \sum_{n=0}^{\infty} P(T_{n+1} - T_n = k,\, T_n \leq t < T_{n+1} \,|\, Z_n = e_i) \\
&\quad = \sum_{n=0}^{\infty} \mathbf{E}[\, \mathrm{I}(T_{n+1} - T_n = k)\, \mathrm{I}(T_n \leq t < T_{n+1}) \,|\, Z_n = e_i] \\
&\quad = \sum_{n=0}^{\infty} \mathbf{E}[\, \mathrm{I}(T_{n+1} - T_n = k)\, \mathrm{I}(T_n \leq t < T_n + k) \,|\, Z_n = e_i] \\
&\quad = \sum_{n=0}^{\infty} \mathbf{E}[\, \mathbf{E}[\, \mathrm{I}(T_{n+1} - T_n = k)\, \mathrm{I}(T_n \leq t < T_n + k) \,|\, Z_n,\, T_n\,] \,|\, Z_n = e_i] \\
&\quad = \sum_{n=0}^{\infty} \mathbf{E}[\, \mathrm{I}(T_n \leq t < T_n + k)\, \mathbf{E}[\, \mathrm{I}(T_{n+1} - T_n = k) \,|\, Z_n,\, T_n\,] \,|\, Z_n = e_i] \\
&\quad = \sum_{n=0}^{\infty} \mathbf{E}[\, \mathrm{I}(T_n \leq t < T_n + k)\, \mathbf{E}[\, \mathrm{I}(T_{n+1} - T_n = k) \,|\, Z_n\,] \,|\, Z_n = e_i] \\
&\quad = \sum_{n=0}^{\infty} \mathbf{E}[\, \mathrm{I}(T_n \leq t < T_n + k)\, \mathbf{E}[\, \mathrm{I}(T_{n+1} - T_n = k) \,|\, Z_n = e_i\,] \,|\, Z_n = e_i] \\
&\quad = p_i(k) \sum_{n=0}^{\infty} \mathbf{E}[\, \mathrm{I}(T_n \leq t < T_n + k) \,|\, Z_n = e_i] \\
&\quad = p_i(k) \sum_{n=0}^{\infty} \mathbf{E}[\, \mathrm{I}(T_n \leq t < T_n + k) \,|\, X_t = e_i] = p_i(k)
\end{aligned}
$$

as

$$\Omega = \bigcup_{n=0}^{\infty} \{T_n \le t < T_n + h_t(X_t)\} = \bigcup_{n=0}^{\infty} \{T_n \le t \le T_n + h_t(X_t) - 1\}$$

by the definition of $h_t(X_t)$.

Calculating the denominator,

$$\begin{aligned} P(h_t(X_t) = k \mid X_t = e_i) &= \sum_{n=0}^{\infty} P(h_t(X_t) = k,\, T_n \le t < T_{n+1} \mid X_t = e_i) \\ &= \sum_{n=0}^{\infty} P(T_{n+1} - T_n \ge k,\, T_n \le t < T_{n+1} \mid Z_n = e_i) \\ &= \sum_{l=k}^{\infty} p_i(l) = F_i(k) \end{aligned}$$

using the same arguments as above. The lemma is proved. □

Corollary 8.4 *If $X_t = e_i$ and $P(h_t(X_t) = k | X_t = e_i) > 0$, then*

$$P(X_{t+1} = e_i \mid X_t = e_i,\, h_t(X_t) = k) = \frac{F_i(k+1)}{F_i(k)}.$$

In the following theorem, we use the notation

$$\frac{\mathbf{p}(k)}{\mathbf{F}(k)} = \left(\frac{p_1(k)}{F_1(k)}, \ldots, \frac{p_N(k)}{F_N(k)}\right)$$

and

$$\frac{\mathbf{F}(k+1)}{\mathbf{F}(k)} = \left(\frac{F_1(k+1)}{F_1(k)}, \ldots, \frac{F_N(k+1)}{F_N(k)}\right).$$

Theorem 8.5 *For all $t \ge 0$,*

$$\mathbf{E}\left[X_{t+1} \mid \mathcal{F}_t^X\right] = B_t(X_t)\, X_t \in \mathbb{R}^N,$$

where

$$B_t(X_t) = A(h_t(X_t)) \left\langle X_t, \frac{\mathbf{p}(h_t(X_t))}{\mathbf{F}(h_t(X_t))} \right\rangle + \left\langle X_t, \frac{\mathbf{F}(h_t(X_t)+1)}{\mathbf{F}(h_t(X_t))} \right\rangle \mathrm{I},$$

and

$$\begin{aligned} A(k)_{ji} &= P(Z_{n+1} = e_j \mid Z_n = e_i, T_{n+1} - T_n = k) \quad \textit{for } j \ne i \\ &= \frac{P(Z_{n+1} = e_j, T_{n+1} - T_n = k \mid Z_n = e_i)}{\mathrm{P}(T_{n+1} - T_n = k \mid Z_n = e_i)} = \frac{q(e_j,\, e_i,\, k)}{q(e_i,\, k)} \end{aligned}$$

where

$$q(e_i,\, k) = \sum_{j=1}^{N} q(e_j,\, e_i.\, k)\,.$$

Proof We have from Theorem 8.2, that if $h_t(X_t) = k$ and $j \neq i$, then a jump has occurred at $t+1$, so using Theorem 8.2 and

$$\begin{aligned} P\left(X_{t+1} = e_j \,|\, \mathcal{F}_{t-1}^X, X_t = e_i\right) &= \mathrm{P}\left(X_{t+1} = e_j \,|\, X_t = e_i,\, h_t(X_t) = k\right) \\ &= A(k)_{ji}, \end{aligned}$$

the result follows. □

We have $\{T_n\}$ for $T_n \in \{0, 1, 2, 3, \ldots\}$ with $T_0 = 0$ and $T_{n+1} - T_n \geq 1$ for each n and these could be random times. We also have marks $\{Z_n\}$ where $Z_n \in \{e_1, e_2, \ldots, e_N\} \subseteq \mathbb{R}^N$. Then, $\{(T_n, Z_n) : n = 0, 1, 2, \ldots\}$ is called a renewal sequence process if the following is true:

$$P(Z_{n+1} = e_j, T_{n+1} - T_n = m \,|\, Z_n = e_i) = q(e_j, e_i, m),$$

which does not depend on n.

We then have a semi-Markov chain

$$\{X_t : t = 0, 1, 2, \ldots\}$$

where $X_t = Z_n$ and $T_n \leq t < T_{n+1}$.

Lemma 8.6 *A Markov chain is a semi-Markov chain.*

Proof Let us suppose that the state space of the Markov chain is $\mathcal{S} = \{e_1, e_2, \ldots, e_N\}$ as we used above.

Let $\{X_n\}$ be a Markov chain with transition probability matrix $A = a_{ji}$. This means that

$$a_{ji} = P(X_{t+1} = e_j \,|\, X_t = e_i)\,.$$

We put $T_0 = 0$ and $Z_0 = X_0$.

We then set

$$T_1 = \min\{n \geq T_0,\, X_n \neq Z_0\}$$

and $Z_1 = X_{T_1}$. We continue in this way. If (T_k, Z_k) is determined, then

$$T_{k+1} = \min\{n \geq T_k,\, X_n \neq Z_k\}$$

and $Z_{k+1} = X_{T_{k+1}}$. Then by definition $Z_{n+1} \neq Z_n$ for all $n \geq 0$. From the previous discussion we can show, for $i \neq j$, that

$$P(Z_{n+1} = e_j, T_{n+1} - T_n = m \,|\, Z_n = e_i) = (1 - a_{ii})a_{ii}^{m-1}$$

which is independent of n. For this we use the idea

$$
\begin{aligned}
&P(Z_{n+1} = e_j, T_{n+1} - T_n = m | Z_n = e_i) \\
&= \sum_k P(Z_{n+1} = e_j, T_{n+1} - T_n = m | Z_n = e_i, T_n = k) P(T_n = k | Z_n = e_i) \\
&= (1 - a_{ii}) a_{ii}^{m-1} \sum_k P(T_n = k | Z_n = e_i) \\
&= (1 - a_{ii}) a_{ii}^{m-1} .
\end{aligned}
$$

This is a geometric distribution and it does not depend on j. □

Çinlar (1975), page 316, provides a result under which a semi-Markov chain is a Markov chain.

Lemma 8.7 *Without lost of generality, we may assume $Z_{n+1} \neq Z_n$ for $n \geq 0$.*

Remark 8.8 Both Çinlar (1975) and Howard (1971) allow $Z_{n+1} = Z_n$ and this is called virtual change of state.

Proof We let $T'_0 = T_0 = 0$ and $Z'_0 = Z_0$. Then

$$T'_1 = \inf \{ n \geq T'_0 : Z_n \neq Z'_0 \} .$$

If we let $Z'_1 = Z_{T'_1}$, then

$$T'_2 = \inf \{ n \geq T'_1 : Z_n \neq Z'_1 \}$$

and $Z'_2 = Z_{T'_2}$ etc.

One can check that $\{(T'_n, Z'_n\}$ is a renewal process and clearly

$$X_t = Z'_n \text{ where } T'_n \leq t < T'_{n+1} \text{ for all } n \geq 0$$

and well as the original

$$X_t = Z_n \text{ where } T_n \leq t < T_{n+1} \text{ for all } n \geq 0.$$ □

From now on, we assume that $Z_{n+1} \neq Z_n$ for all $n \geq 0$ in the renewal processes.

Recall the notations

$$Z_{0:n} = \{Z_0, Z_1, \ldots, Z_n\}.$$

Then we can write

$$
\begin{aligned}
&P(Z_{n+1} = e_j, T_{n+1} - T_n = m \,|\, Z_{0:n}, T_{0:n}) \\
&\quad = P(Z_{n+1} = e_j, T_{n+1} - T_n = m \,|\, Z_n) .
\end{aligned}
$$

Lemma 8.9 *Given $Z_{0:n-1}$, then $T_1 - T_0, T_2 - T_1, \ldots, T_n - T_{n-1}$ are independent.*

Proof

$$
\begin{aligned}
&P(T_1 - T_0 = m_1, T_2 - T_1 = m_2, \ldots, T_n - T_{n-1} = m_n \mid Z_{0:n-1}) \\
&\quad = \mathbf{E}\left[\prod_{l=1}^{n} I(T_l - T_{l-1} = m_l) \,\middle|\, Z_{0:n-1}\right] \\
&\quad = \mathbf{E}\left[\mathbf{E}\left[\prod_{l=1}^{n} I(T_l - T_{l-1} = m_l) \,\middle|\, Z_{0:n-1}, T_{0:n-1}\right] \,\middle|\, Z_{0:n-1}\right] \\
&\quad = \mathbf{E}\left[\prod_{l=1}^{n-1} I(T_l - T_{l-1} = m_l) \times \right. \\
&\qquad\qquad \left. \times\, \mathbf{E}\left[I(T_n - T_{n-1} = m_n) \mid Z_{0:\,n-1}, T_{0:n-1}\right] \,\middle|\, Z_{0:\,n}\right] \\
&\quad = \mathbf{E}\left[\prod_{l=1}^{n-1} I(T_l - T_{l-1} = m_l) \times \right. \\
&\qquad\qquad \left. \times\, \mathbf{E}\left[I(T_n - T_{n-1} = m_n) \mid Z_{n-1}\right] \,\middle|\, Z_{0:\,n-1}\right] \\
&\quad = \mathbf{E}\left[I(T_n - T_{n-1} = m_n) \mid Z_{n-1}\right] \times \\
&\qquad\qquad \times\, \mathbf{E}\left[\prod_{l=1}^{n-1} I(T_l - T_{l-1} = m_l) \,\middle|\, Z_{0:\,n-2}\right] \\
&\quad = \prod_{l=1}^{n} P(T_l - T_{l-1} = m_l \mid Z_{l-1}) \,.
\end{aligned}
$$

and the result is proved. □

Application Note

$$
\begin{aligned}
&P(T_2 - T_0 = m \mid Z_0 = e_i, Z_1 = e_i) \\
&\quad = \sum_{k=1}^{m-1} P\left(T_1 - T_0 = k, T_2 - T_1 = m - k \mid Z_0 = e_i, Z_1 = e_i\right) \\
&\quad = \sum_{k=1}^{m-1} P(T_1 - T_0 = k \mid Z_0 = e_i)\, P(T_2 - T_1 = m - k \mid Z_1 = e_i).
\end{aligned}
$$

8.4 Exercises

Exercise 8.1 Prove the final inequality in the proof of Theorem 8.2.

Exercise 8.2 Show that $\{(T'_n, Z'_n)\}$ defined in Lemma 8.7 is a renewal process.

9

Hidden Semi-Markov Models

9.1 Introduction

The chapter first presents a construction of a semi-Markov process X on its canonical probability space, the space of all sequences of elements of the state space. It is then supposed that the semi-Markov chain is not directly observed but that there is a second finite state process Y whose transitions depend on the state of the hidden process X.

In Chapter 8 we described a process $\{(T_n, Z_n)\}$ with

$$Z_{n+1} \neq Z_n, \quad n \geq 0.$$

We also had

$$\begin{aligned} &q(e_j, e_i, m) \\ &\quad = P(Z_{n+1} = e_j, T_{n+1} - T_n = m \mid Z_n = e_i) \\ &\quad = P(T_{n+1} - T_n = m \mid Z_{n+1} = e_j, Z_n = e_i)\, P(Z_{n+1} = e_j \mid Z_n = e_i) \\ &\quad = \tau(e_j, e_i, m)\, p_{j\,i}. \end{aligned} \tag{9.1}$$

With this decomposition, the parameters of the model are $\{\tau(e_j, e_i, m),$ with $j \neq i$, $m = 1, 2, 3, \ldots\}$ and $\{p_{j\,i}, j \neq i\}$ and these are to be estimated.

An alternative decomposition is

$$\begin{aligned} &q(e_j, e_i, m) \\ &= P(Z_{n+1} = e_j \mid T_{n+1} - T_n = m,\, Z_n = e_i)\, P(T_{n+1} - T_n = m \mid Z_n = e_i) \\ &= A_{j\,i}(m)\, p_i(m)\,. \end{aligned}$$

With this decomposition, the parameters of the model that are to be estimated are $\{A_{j\,i}(m), j \neq i, m = 1, 2, 3, \ldots\}$ and $\{p_i(m), m = 1, 2, 3, \ldots\}$. In fact we shall estimate this second decomposition.

The work of Ferguson (1980), Burge (1997), Burge and Karlin (1997),

Bulla (2006), Bulla and Bulla (2006), Bulla et al. (2010), Guédon and Cocozza-Thivent (1990) and others have used the second specification but with $A_{j\,i}$ not depending on m. In these works the model is simulated by first selecting $X_0 = Z_0$ according to an initial distribution $\{\pi_i(0),\, i = 1, 2, \ldots, N\}$. If $X_0 = e_i$, then a duration $T_1 - T_0$ is selected from the distribution given by $\{p_i(m),\, m \geq 1\}$ and then a change of state according to the distribution $\{A_{j\,i},\, j = 1, 2, \ldots, N,\, j \neq i\}$ and so on. The specification of these authors is equivalent to assuming that $\tau_{j\,i}(m)$ in the first formulation does not depend on j. When applying this restricted formulation, it is necessary to determine if the model is rich enough to model the application.

Some authors give parametric forms to $\{p_i(m)\}$. This reduces the number of parameters of the model, but a suitable parametric model will need to be justified for each application. See Levinson (1986a,b), Ramesh and Wilpon (1992) Guédon (1992, 1999, 2003, 2007) and Guédon and Cocozza-Thivent (1990). A good review of various approaches is Yu (1986) but we shall use different notation and provide some alternative estimates.

Recall that a semi-Markov process $\{X_t : t = 0, 1, 2, \ldots\}$ is expressed in terms of a renewal process by

$$X_t = Z_n \quad \text{if } T_n \leq t < T_{n+1}.$$

We have established the key result:

$$P\left(X_{t+1} = e_j \,|\, \mathcal{F}_t^X\right) = P\left(X_{t+1} = e_j \,|\, X_t, h_t(X_t)\right),$$

where, if $X_t = e_i$, say, we set

$$h_t(X_t) = k$$

when

$$X_t = e_i, X_{t-1} = e_i, \ldots, X_{t-k+1} = e_i, X_{t-k} \neq e_i.$$

Sometimes in what follows, we shall write h_t for $h_t(X_t)$ and note that

$$h_t = t - T_n + 1 \quad \text{if} \quad T_n \leq t < T_{n+1}$$

for $t = 0, 1, 2, \ldots$ It follows from the definition of a semi-Markov chain that $\{(X_t, h_t), t = 0, 1, 2, \ldots\}$ is a Markov chain with state space

$$\{(e_j, d),\, j = 1, 2, \ldots, N,\, d = 1, 2, 3, \ldots\}$$

for which a transition matrix can be written down. Krishnamurthy et al. (1991) use this extended state space as do Yu and Kobayashi in their papers (2003a, 2003b, 2006).

We used the following important quantities for integers $d \geq 1$ in Lemma 8.3:

$$p_i(d) = P(T_{n+1} - T_n = d \,|\, Z_n = e_i) = \sum_{j \neq i} q(e_j, e_i, m),$$

$$F_i(d) = \sum_{l=k}^{\infty} p_i(l) = P(T_{n+1} - T_n \geq d \,|\, Z_n = e_i).$$

We shall also use the formula

$$A_{j\,i}(m) = \frac{q(e_j,\, e_i\; m)}{\sum_{j \neq i} q(e_j,\, e_i\; m)}$$

With these notations we now establish

Lemma 9.1 *For $j \neq i$ and $d \geq 1$:*

(a) $P(X_{t+1} = e_j,\, h_t = d \,|\, X_t = e_i,\, X_{t+1} \neq e_i) = q(e_j,\, e_i,\, d)$;
(b) $P(X_{t+1} = e_j \,|\, X_t = e_i,\, X_{t+1} \neq e_i) = p(e_j,\, e_i) = \sum_{d \geq 1} q(e_j,\, e_i,\, d)$;
(c) $P(X_{t+1} = e_j \,|\, X_t = e_i,\, X_{t+1} \neq e_i,\, h_t = d) = A_{j\,i}(d)$;
(d) $P(X_{t+1} = e_i \,|\, X_t = e_i,\, h_t = d) = \dfrac{F_i(d+1)}{F_i(d)}$;
(e) $P(X_{t+1} \neq e_i \,|\, X_t = e_i,\, h_t = d) = \dfrac{p_i(d)}{F_i(d)}$.

Proof The proof of (a) uses the same technique as in Theorem 8.2. Equation (b) follows from (a). Equation (c) again uses the techniques of Theorem 8.2. Equations (d) and (e) are Lemma 8.3 and its Corollary 8.4. □

Corollary 9.2 *For any $1 \leq j,\, i \leq N$ and $d \geq 1$,*

$$P(X_{t+1} = e_j \,|\, X_t = e_i,\, h_t = d) = A_{j\,i}(d) \cdot \frac{p_i(d)}{F_i(d)} + \delta_{j\,i} \cdot \frac{F_i(d+1)}{F_i(d)}$$

where $\delta_{j\,i} = 1$ if $i = j$ and $\delta_{j\,i} = 0$, otherwise.

Proof We have

$$\begin{aligned}
&P(X_{t+1} = e_j \,|\, X_t = e_i,\, h_t = d) \\
&= P(X_{t+1} = e_j \,|\, X_t = e_i,\, X_{t+1} \neq e_i,\, h_t = d) \\
&\qquad \cdot P(X_{t+1} \neq e_i \,|\, X_t = e_i,\, h_t = d) \\
&\quad + P(X_{t+1} = e_j \,|\, X_t = e_i,\, X_{t+1} = e_i,\, h_t = d) \\
&\qquad \cdot P(X_{t+1} = e_i \,|\, X_t = e_i,\, h_t = d) \\
&= A_{j\,i}(d) \cdot \frac{p_i(d)}{F_i(d)} + \delta_{j\,i} \cdot \frac{F_i(d+1)}{F_i(d)}
\end{aligned}$$

and the result is proved. □

9.2 A semi-martingale representation for a semi-Markov chain

As above, $X = \{x_k,\, k = 0, 1, 2, \ldots\}$ is a semi-Markov chain with jump times $0 < T_1 < T_2 < T_3 < \cdots$.

Write $X_{T_n} = Z_n \in S = \{e_1, e_2, \ldots, e_N\}$ and

$$\mathcal{F} = \sigma\{Z_k : k \leq t\}.$$

The semi-Markov property states that

$$\begin{aligned}
&P\big(Z_{n+1} = e_j,\, T_{n+1} - T_n = m | \mathcal{F}_{T_n}\big) \\
&\quad = P\big(Z_{n+1} = e_j,\, T_{n+1} - T_n = m | Z_n = e_i\big) \\
&\quad = P\big(T_{n+1} - T_n = m | Z_{n+1} = e_j,\, Z_n = e_i\big) P\big(Z_{n+1} = e_j | Z_n = e_i\big) \\
&\quad = q(e_j, e_i, m) \\
&\quad = \tau_{ji}(m) p_{ji}
\end{aligned}$$

where

$$P\big(T_{n+1} - T_n = m | Z_{n+1} = e_j,\, Z_n = e_i\big) = \tau_{ji}(m)$$

and

$$P\big(Z_{n+1} = e_j | Z_n - e_i\big) = p_{ji}.$$

In this section, we suppose $\tau_{ji}(m)$ does not depend on e_j so:

$$\begin{aligned}
&P\big(T_{n+1} - T_n = m | Z_{k+1} = e_j,\, Z_n = e_i\big) \\
&\qquad = P\big(T_{n+1} - T_n = m | Z_n = e_i\big) = p_i(m), \qquad \text{say.}
\end{aligned}$$

Since the process is homogeneous this means these probabilities $p_i(m)$ are independent of n.

Write

$$\begin{aligned}
G_i(k) &= \sum_{m=1}^{k-1} p_i(m) = P(T_{n+1} - T_n < k | Z_n = e_i) \\
F_i(k) &= P(T_{n+1} - T_n \geq k | Z_n = e_i) = 1 - G_i(k) \\
F_i(k, j) &= P(T_{n+1} - T_n \geq k,\, Z_{n+1} = e_j | X_n = e_i) \\
&= F_i(k) p_{ji}\,.
\end{aligned}$$

Consider the processes:

$$p_i^n(k,j) = I_{T_n+k \geq T_{n+1}} I(Z_{n+1}=e_j) I(Z_n = e_i)$$
$$\widetilde{p}_i^n(k,j) = \sum_{T_n < T_n+m \leq T_n+k \wedge T_{n+1}} \frac{P_i(m)}{F_i(m)} p_{ji}$$
$$q_i^n(k,j) = p_i^n(k,j) - \widetilde{p}_i^n(k,j).$$

Theorem 9.3 *Suppose $T_n < T_n + d \leq T_n + k$. Then*

$$E\big[q_i^n(k,j) | \mathcal{F}_d\big] = q_i^n(d,j)$$

for all appropriate n, i, j.

Proof

$$E\big[p_i^n(k,j) - p_i^n(d,j) | \mathcal{F}_d\big] = I_{T_{n+1} > T_n+d} \left(\frac{G_i(k+1) - G_i(d+1)}{F_i(d+1)} \right) p_{ji},$$

Also,

$$\begin{aligned} & E[\widetilde{p}_i^n(k,j) - \widetilde{p}_i^n(d,j) | \mathcal{F}_d] \\ &= I_{T_{n+1} > T_n+d} E\Big[\sum_{T_n+d < T_n+m \leq T_n+k \wedge T_{n+1}} - \left(\frac{p_i(m)}{F_i(m)} \right) p_{ji} | \mathcal{F}_d \Big] \\ &= I_{T_{n+1} > T_n+d} \frac{p_{ji}}{F_i(d+1)} \Big[F_i(k+1) \sum_{T_n+d < T_n+m \leq T_n+k} \left(\frac{p_i(m)}{F_i(m)} \right) \\ &\quad + \sum_{T_n+d < T_n+r \leq T_n+k} \Big(\sum_{T_n+d < T_n+m \leq T_n+r} \frac{p_i(m)}{F_i(m)} \Big) p_i(r) \Big]. \end{aligned}$$

Interchanging the order in the last double sum gives

$$\begin{aligned} & \sum_{T_n+d < T_n+r \leq T_n+k} \Big(\sum_{T_n+d < T_n+m \leq T_n+r} \frac{p_i(m)}{F_i(m)} \Big) p_i(r) \\ &= \sum_{T_n+d < T_n+m \leq T_n+k} \Big(\sum_{T_n+m \leq T_n+r \leq T_n+k} p_i(r) \Big) \frac{p_i(m)}{F_i(m)} \\ &= \sum_{T_n+d < T_n+m \leq T_n+k} \big(F_i(m) - F_i(k+1) \big) \frac{p_i(m)}{F_i(m)} \\ &= \sum_{T_n+d < T_n+m \leq T_n+k} p_i(m) - F_i(k+1) \sum_{T_n+d < T_n+m \leq T_n+k} \frac{p_i(m)}{F_i(m)} \\ &= G_i(k+1) - G_i(d+1) - F_i(k+1) \sum_{T_n+d < T_n+m \leq T_n+k} \frac{p_i(m)}{F_i(m)} \end{aligned}$$

Therefore

$$
\begin{aligned}
&E[\widehat{p}_i^n(k,j) - \widetilde{p}_i^n(d,j)|\mathcal{F}_d] \\
&= I_{T_{n+1}<T_n+d}\,\frac{p_{ji}}{F_i(d+1)}\left(G_i(k+1) - G_i(d+1)\right) \\
&= E[p_i^n(k,j) - p_i^n(d,j)|\mathcal{F}_d].
\end{aligned}
$$

so

$$
E[q_i^n(k,j)|\mathcal{F}_d] = q_i^n(d,j)\,. \qquad \square
$$

Corollary 9.4 *Write $Q(m) = \big(Q_{ji}(m),\, 1 \le i,\, j \le N\big)$ for the matrix with entries*

$$
Q_{ji}(m) = \frac{p_i(m)}{F_i(m)}\,p_{ji}\,.
$$

Then for $T_n + k \ge 1$:

$$
q^n(T_n+k) := I_{T_n+k\ge T_{n+1}}\,Z_{n+1} - \sum_{T_n<T_n+m\le T_n+k\wedge T_{n+1}} Q_m Z_n \in \mathbb{R}^N
$$

is an $\{\mathcal{F}_k\}$-martingale.

Proof With

$$
p_i^n(k,j) = I_{T_n+k\ge T_{n+1}}\,\langle Z_{n+1}, e_j\rangle\,\langle Z_n, e_i\rangle
$$

and

$$
\widetilde{p}_i(k,j) = \sum_{T_n<T_n+m\le T_n+k\wedge T_{n+1}} \frac{p_i(m)}{F_i(m)}\,p_{ji}
$$

we have shown that $p_i^n(k,j) - \widetilde{p}_i(k,j)$ is a martingale. However,

$$
\begin{aligned}
I_{T_n+k\ge T_{n+1}} Z_{n+1} &:= p^n(k) \\
&= I_{T_n+k\ge T_{n+1}}\left(\sum_{i=1}^{n}\langle Z_n, e_i\rangle\right)\left(\sum_{j=1}^{N}\langle Z_{n+1}, e_j\rangle\right) e_j
\end{aligned}
$$

so with

$$
\begin{aligned}
\widetilde{p}^n(k) &= \sum_{j=1}^{N}\sum_{i=1}^{N}\sum_{T_n<T_n+m\le T_n+k\wedge T_{n+1}} \frac{p_i(m)}{F_i(m)}\,p_{ji}\,\langle Z_n, e_i\rangle\, e_j \\
&= \sum_{T_n<Tn+m\le T_n+k\wedge T_{n+1}} Q(m) Z_n\,,
\end{aligned}
$$

$p^n(k) - \widetilde{p}^k(k)$ is a martingale. $\square$

Corollary 9.5 *Write*

$$q(t) = \sum_n I(T_n \leq T_n + t) q^n(t).$$

Then $q(t)$ is an $\{\mathcal{F}_t\}$-martingale.

Proof Suppose $T_m < T_n + s \leq T_{m+1} \leq T_n < T_n + t$. Note $p^n(t)$ and $\widetilde{p}^n(t)$ are only defined for $t > T_n + 1$. With $q^n(t) = p^n(t) - \widetilde{p}^n(t)$ and $t \geq T_n + 1$.

$$\begin{aligned} E[q^n(t)|\mathcal{F}_{T_n+1}] &= q^n(T_n + 1) \\ &= p^n(T_n + 1) - \widetilde{p}(T_n + 1) \\ &= I_{T_n+1=T_{n+1}} Z_{n+1} - Q(T_n + 1) Z_n \,. \end{aligned}$$

For $i \neq j$,

$$Q_{ji}(T_n + 1) = \frac{p_i(T_n + 1)}{p_i(T_n + 1)} \, p_{ji}$$

so

$$\begin{aligned} Q(T_n + 1) &= (p_{ji}, \, 1 \leq i, \, j \leq N) \\ &= \Pi \quad \text{say}. \end{aligned}$$

Also,

$$E[I_{T_{n+1}} Z_{n+1} | \mathcal{F}_{T_n}] = \Pi \, Z_n$$

so

$$E[q^n(t)|\mathcal{F}_{T_n}] = E[q^n(T_n + 1)|\mathcal{F}_{T_n}] = 0 \in \mathbb{R}^N. \qquad \square$$

Corollary 9.6 *The martingale*

$$\sum_{n \geq 0} q^n(t) = q(t)$$

is associated with the semi-Markov chain $X = \{X_t, \, t = 0, 1, 2, \ldots\}$. Write

$$\widetilde{p}(t) = \sum_{n \geq 0} \widetilde{p}^n(t).$$

Then X has the semi-martingale representation

$$X_t = X_0 + q(t) + \widetilde{p}(t) \in \mathbb{R}^N.$$

9.3 Construction of the semi-Markov model

We start with the usual reference probability space $(\Omega, \mathcal{F}, \overline{P})$ and let

$$\overline{\lambda}_0 \equiv N\langle \pi(0), X_0 \rangle,$$

and

$$\overline{\lambda}_{l+1} \equiv N\langle B_l(X_l)\, X_l, X_{l+1} \rangle, \quad l \geq 0.$$

We define

$$\overline{\Lambda}_k \equiv \prod_{l=0}^{k} \overline{\lambda}_l.$$

Note that $\pi_i(0)$ will be $P(X_0 = e_i)$.

Lemma 9.7 *We have*

$$\mathbf{E}\left[\overline{\lambda}_l \,\middle|\, \mathcal{F}_{l-1}^X\right] = 1 \textit{ for } l \geq 1 \textit{ and } \mathbf{E}\left[\overline{\lambda}_0\right] = 1\,.$$

Proof This is straightforward and we omit the details. □

We next define P on $(\Omega, \mathcal{F})$ by

$$\left.\frac{dP}{d\overline{P}}\right|_{\mathcal{F}_k^X} = \overline{\Lambda}_k.$$

This P exists, as in Chapter 1 and, under P,

$$\mathbf{E}\left[X_{t+1} \,|\, \mathcal{F}_t^X\right] = B_t(X_t)\, X_t$$

using the notation of Chapter 8. Conversely, if $E[X_{t+1}|\mathcal{F}_t^X] = B_t(X_t)X_t$ for $t = 0, 1, 2, \ldots$, then $\{X_t,\, t \geq 0\}$ is a semi-Markov chain.

9.4 Hidden semi-Markov models

For hidden semi-Markov models, suppose we have a chain of observations given by

$$\{Y_k \,:\, k = 0, 1, 2, \ldots\}$$

where $Y_k \in \{f_1, \ldots, f_M\} \subset \mathbb{R}^M$. for each k.

We use the notation

$$\mathcal{Y}_k = \sigma\{Y_0, Y_1, \ldots, Y_k\}$$
$$\mathcal{F}_k = \sigma\{X_0, X_1, \ldots, X_k\}$$
$$\mathcal{G}_k = \mathcal{Y}_k \vee \mathcal{F}_k$$

as before.

We shall consider two possible specifications. We use the terminology given in Barbu and Limnios (2008).

(a) The homogeneous SM–M0 model This is specified by the property

$$P(Y_k = f_r \,|\, \mathcal{G}_{k-1} \vee \mathcal{F}_k) = P(Y_k = f_r \,|\, X_k)$$

and we define

$$C_{ri} \equiv P(Y_k = f_r \,|\, X_k = e_i),$$

which does not depend on k, and is termed homogeneous.

(b) The homogeneous SM–M1 model This is specified by the property

$$P(Y_k = f_r \,|\, \mathcal{Y}_{k-1} \vee \mathcal{F}_k) = P(Y_k = f_r \,|\, Y_{k-1}, X_k).$$

This means that $\{Y_k\}$ is a first-order (conditional or modulated) Markov chain which belongs to class **M1** in the notation of Barbu and Limnios. In this case, we define

$$P(Y_k = f_r \,|\, Y_{k-1} = f_s, X_k = e_i) = C^i_{rs}$$

as before.

Using the representation of higher-order chains given in Section 2, where higher-order chains can be expressed as first order chains with an extended state space, the discussion we provide for this model can be extended to the so-called **homogeneous SM–Mk model** for $k \geq 1$.

Constructions (a) With the SM–M0 model, define

$$\overline{\lambda}_0 = N\, M\, \langle \pi(0), X_0 \rangle \, \langle DX_0, Y_o \rangle,$$

where

$$P(Y_0 = f_r \,|\, X_0 = e_i) \equiv D_{ri}$$

and

$$\overline{\lambda}_{l+1} = N\, M\, \langle B_l(X_l)\, X_l, X_{l+1} \rangle \, \langle CX_{l+1}, Y_{l+1} \rangle$$

for $l \geq 0$.

We also define

$$\overline{\Lambda}_k = \prod_{l=0}^{k} \overline{\lambda}_l$$

for $k \geq 0$.

Lemma 9.8 *For $l \geq 0$, we have*

$$\mathbf{E}\left[\overline{\lambda}_0\right] = 1 \text{ and } \mathbf{E}\left[\overline{\lambda}_{l+1} \,|\, \mathcal{G}_l\right] = 1.$$

Proof This is straightforward and we omit the details. □

We now define P on the measure space $(\Omega, \mathcal{F})$ as introduced in Chapter 1 so that

$$\left.\frac{dP}{d\overline{P}}\right|_{\mathcal{G}_k} = \overline{\Lambda}_k,$$

for $k \geq 0$. Under this P, X and Y satisfy the correct model dynamics.

(b) For the SM–M1 Model (including the SM–Mk models), we define

$$\overline{\lambda}_0 = N\, M\, \langle \pi(0), X_0 \rangle \, \langle D X_0, Y_0 \rangle$$

and

$$\overline{\lambda}_{l+1} = N\, M\, \langle B_l(X_l)\, X_l, X_{l+1} \rangle \, \langle C(X_{l+1})\, Y_l, Y_{l+1} \rangle$$

for $l \geq 0$, where

$$C(X_{l+1}) = \sum_{i=1}^{N} C^i \langle X_{l+1}, e_i \rangle \, .$$

Lemma 9.9 *We have*

$$\mathbf{E}\left[\overline{\lambda}_0\right] = 1 \text{ and } \mathbf{E}\left[\overline{\lambda}_{l+1} \,|\, \mathcal{G}_l\right] = 1$$

for $l \geq 0$.

We again define P on the measure space $(\Omega, \mathcal{F})$ introduced in Chapter 1 so that

$$\left.\frac{dP}{d\overline{P}}\right|_{\mathcal{G}_k} = \overline{\Lambda}_k,$$

for $k \geq 0$. Under this P, X and Y satisfy the correct model dynamics. We omit the details, but arguments like those used in Chapter 1 are employed.

We shall now assume that we have observations $\{y_0, y_1, \dots, y_n\}$ of $\{Y_0, Y_1, \dots, Y_n\}$ and we often write $Y_{0:n} = y_{0:n}$ to express this.

We shall also focus our proofs on the homogeneous SM–M1 model and provide without proof the results for the homogeneous SM–M0 model which could be established in a similar (and often simpler) way.

Filters Given a sequence of observations $y_{0:k} \in \mathcal{S}^{k+1}$ we wish to compute (using Koski's notation)

$$\hat{\pi}_i(k \mid k) = P_\theta(X_k = e_i \mid Y_{0:k} = y_{0:k}),$$

where θ represents the parameters of the model. These parameters are those of the the semi-Markov chain and the emissions matrix C in its various specifications.

These quantities are given in terms of un-normalized quantities $\{\alpha_i(k)\}$ by

$$\hat{\pi}_i(k \mid k) = \frac{\alpha_i(k)}{\sum_{j=1}^N \alpha_j(k)}$$

where

$$\alpha_i(k) = P_\theta(Y_{0:k} = y_{0:k},\, X_k = e_i)$$

from which we have

$$P_\theta(X_k = e_i \mid Y_{0:k} = y_{0:k}) = \frac{P_\theta(X_k = e_i, Y_{0:k} = y_{0:k})}{P_\theta(Y_{0:k} = y_{0:k})}.$$

We show that $\{\alpha_i(k)\}$ can be found via a forward recurrence.

In order to obtain this recurrence, we introduce

$$\alpha_{i,\,d}(k) = P_\theta(Y_{0:k} = y_{0:k},\, X_k = e_i,\, h_k = d)$$

for $1 \leq i \leq N,\ 1 \leq d \leq k+1,\ 0 \leq k \leq n$, and note that

$$\alpha_i(k) = \sum_{d=1}^{k+1} \alpha_{i,\,d}(k).$$

It is also convenient to introduce, as in Chapter 9,

$$q^i_{k,\,d} = \overline{\mathbf{E}}[\,\overline{\Lambda}_k \,\langle X_k,\, e_i\rangle \,\mathrm{I}(h_k = d) \mid \mathcal{Y}_k]$$

which are the typical type of quantities used by (Elliott et al., 1995, pages 26, 62). Because of the Doob–Dynkin result, we could write

$$q^i_{k,\,d} = q^i_{k,\,d}(Y_{0:k})$$

and we would see that

$$q^i_{k,\,d}(y_{0:k}) = \overline{\mathbf{E}}[\,\overline{\Lambda}_k \,\langle X_k,\, e_i\rangle \,\mathrm{I}(h_k = d) \mid Y_{0:k} = y_{0:k}]\,.$$

Lemma 9.10 *For $1 \leq i \leq N,\ 1 \leq d \leq k+1$ and $0 \leq k \leq n$, we have*

$$\alpha_{i,\,d}(k) = M^{-k-1}\, q^i_{k,\,d}(y_{0:k}).$$

Proof This result is shown in a similar same way to that of Lemma 7.3 because the explicit form of the $\overline{\Lambda}_k$ is not needed. □

It is more convenient to obtain a recurrence relationship for $\{\alpha_{i,\,d}(k)\}$ using other quantities. It will also be convenient to use the (shorthand) notation

$$\Gamma(j,\,i,\,d) = A_{j\,i}(d)\cdot\frac{p_i(d)}{F_i(d)} + \delta_{j\,i}\cdot\frac{F_i(d+1)}{F_i(d)}$$

for each $1 \le i,\, j \le N$ and $d \ge 1$.

Theorem 9.11 *We have the recurrence relationship*

$$\alpha_{i,\,d}(k) = \langle C(e_i)\, y_{k-1},\, y_k\rangle \sum_{j,\,d'} \Gamma(i,\; j,\, d')\, \alpha_{j,\,d'}(k-1) \tag{9.2}$$

with

$$\alpha_{i,\,d}(0) = \pi_i(0)\,\langle D\, e_i,\, y_0\rangle\, \mathrm{I}(d=1)\,. \tag{9.3}$$

Remark We make some remarks about the summation in (9.2). If $d = 1$, the summation is over all $1 \le d' \le k$ and over $1 \le j \le N$ but with $j \ne i$. If $d > 1$, then $d' = d - 1$ in the sum and $j = i$ only. We can then express (9.2) as

$$\alpha_{i,\,1}(k) = \langle C(e_i)\, y_{k-1},\, y_k\rangle \sum_{j=1}^{N}\sum_{d'=1}^{k} A_{i\,j}(d')\,\frac{p_j(d')}{F_j(d')}\,\alpha_{j,\,d'}(k-1)$$

where we recall that $A_{i\,j}(d') = 0$ when $i = j$. When $d > 1$, equation (9.2) becomes

$$\alpha_{i,\,d}(k) = \langle C(e_i)\, y_{k-1},\, y_k\rangle\, \frac{F_i(d)}{F_i(d-1)}\,\alpha_{i,\,d-1}(k-1)$$

which are expressions that only use the parameters of the model.

We also recall the notations

$$b^i(y_k, y_{k-1}) = \langle C(e_i)\, y_{k-1},\, y_k\rangle$$

for the SM–M1 model and

$$b^i(y_k) = \langle C\, e_i,\, y_k\rangle$$

for the SM–M0 model.

Proof We have

$$\alpha_{i,\,d}(k) = M^{-k-1}\,\overline{\mathbf{E}}[\overline{\Lambda}_k\,\langle X_k,\, e_i\rangle\, \mathrm{I}(h_k = d)\,|\, Y_{0:k} = y_{0:k}]$$

$$= N\,M^{-k}\,\overline{\mathbf{E}}\left[\overline{\Lambda}_{k-1}\,\langle B_{k-1}(X_{k-1})\,X_{k-1},\,X_k\rangle\,\langle C(X_k)\,Y_{k-1},\,Y_k\rangle\right.$$
$$\left.\times\,\langle X_k,\,e_i\rangle\,\mathrm{I}(h_k=d)\,\middle|\,Y_{0:k}=y_{0:k}\right]$$
$$= NM^{-k}\langle C(e_i)\,y_{k-1},\,y_k\rangle\,\overline{\mathbf{E}}\left[\overline{\Lambda}_{k-1}\,\langle B_{k-1}(X_{k-1})\,X_{k-1},\,e_i\rangle\right.$$
$$\left.\times\,\langle X_k,\,e_i\rangle\,\mathrm{I}(h_k=d)\,\middle|\,Y_{0:k}=y_{0:k}\right]$$
$$= NM^{-k}\langle C(e_i)y_{k-1},y_k\rangle\sum_{j=1}^{n}\sum_{d'=1}^{k}\overline{\mathbf{E}}\left[\overline{\Lambda}_{k-1}\,\langle X_{k-1},e_j\rangle\,\mathrm{I}(h_{k-1}=d')\right.$$
$$\left.\times\langle B_{k-1}(X_{k-1})\,X_{k-1},\,e_i\rangle\,\langle X_k,\,e_i\rangle\,\mathrm{I}(h_k=d)\,\middle|\,Y_{0:k}=y_{0:k}\right]$$
$$= N\,M^{-k}\,\langle C(e_i)\,y_{k-1},\,y_k\rangle\,\sum_{j=1}^{n}\sum_{d'=1}^{k}\Gamma(i,\,j,\,d')$$
$$\times\,\overline{\mathbf{E}}\left[\overline{\Lambda}_{k-1}\,\langle X_{k-1},\,e_j\rangle\,\mathrm{I}(h_{k-1}=d')\,\langle X_k,\,e_i\rangle\,\mathrm{I}(h_k=d)\,\middle|\,Y_{0:k}=y_{0:k}\right]$$

and now the quantity

$$N\,\overline{\mathbf{E}}\left[\overline{\Lambda}_{k-1}\,\langle X_{k-1},\,e_j\rangle\,\mathrm{I}(h_{k-1}=d')\,\langle X_k,\,e_i\rangle\,\mathrm{I}(h_k=d)\,\middle|\,Y_{0:k}=y_{0:k}\right]$$

equals

$$\overline{\mathbf{E}}\left[\overline{\Lambda}_{k-1}\,\langle X_{k-1},\,e_j\rangle\,\mathrm{I}(h_{k-1}=d')\,\middle|\,Y_{0:k}=y_{0:k}\right]=M^k\,\alpha_{j,\,d'}(k-1)$$

when $d=1$ and $i\neq j$, zero when $d=1$ and $i=j$ and equals

$$\overline{\mathbf{E}}\left[\overline{\Lambda}_{k-1}\,\langle X_{k-1},\,e_i\rangle\,\mathrm{I}(h_{k-1}=d-1)\,\middle|\,Y_{0:k}=y_{0:k}\right]=M^k\,\alpha_{i,\,d-1}(k-1)$$

when $d>1$ and $j=i$ and zero if $d>1$ and $j\neq i$.

The formula for $\alpha_{i,\,d}(0)$ is straightforward. □

Remark In the case of the SM–M0 model, the recursion is the same but the term $\langle C(e_i)\,y_{k-1},\,y_k\rangle$ is replaced with $\langle C\,e_i,\,y_k\rangle$.

It is awkward to express (9.2) in matrix form, but it can be easily implemented in MATLAB as the indices $1\leq i\leq N$, $0\leq k\leq n$ and $1\leq d\leq n+1$ are bounded, given $Y_{0:n}=y_{0:n}$ as the observations.

Lemma 9.12 (This generalizes Lemma 3.7.)

For any $n\geq 1$ *and* $y_l\in\{f_1,f_2,\ldots,f_M\}$ *for* $0\leq l\leq n$,

$$P_\theta(Y_{0:n}=y_{0:n}\,|\,\mathcal{F}_n)=\langle D\,X_0,\,y_0\rangle\prod_{l=1}^{n}\langle C(X_l)\,y_{l-1},\,y_l\rangle$$

for the hidden NM–M1 *model, and for the hidden* NM–M0 *model, we have*

$$P_\theta(Y_{0:n}=y_{0:n}\,|\,\mathcal{F}_n)=\prod_{l=0}^{n}\langle C\,X_l,\,y_l\rangle\,.$$

Proof We provide details for the first case. The left-hand side of this equality can be computed via Bayes' Conditional Expectation Lemma:

$$P_\theta(Y_{0:n} = y_{0:n} \mid \mathcal{F}_n) = \mathbf{E}_\theta\left[\prod_{l=0}^{n}\langle Y_l,\, y_l\rangle \,\middle|\, \mathcal{F}_n\right]$$
$$= \frac{\overline{\mathbf{E}}\left[\overline{\Lambda}_n \prod_{l=0}^{n}\langle Y_l,\, y_l\rangle \mid \mathcal{F}_n\right]}{\overline{\mathbf{E}}\left[\overline{\Lambda}_n \mid \mathcal{F}_n\right]}.$$

We now concentrate on the numerator as the expression for the denominator then follows:

$$\overline{\mathbf{E}}\left[\overline{\Lambda}_n \prod_{l=0}^{n}\langle Y_l,\, y_l\rangle \,\middle|\, \mathcal{F}_n\right]$$
$$= \mathbf{E}\Bigg[M^{n+1}\, N^{n+1}\,\langle \pi(0),\, X_0\rangle\,\langle DX_0,\, Y_0\rangle \prod_{l=1}^{n}\langle AX_{l-1},\, X_l\rangle$$
$$\times \prod_{l=1}^{n}\langle C(X_l)Y_{l-1},\, Y_l\rangle \prod_{l=0}^{n}\langle Y_l,\, y_l\rangle \,\Bigg|\, \mathcal{F}_n\Bigg]$$

$$= M^{n+1}\, N^{n+1}\,\langle \pi(0),\, X_0\rangle\,\langle DX_0,\, y_0\rangle$$
$$\times \prod_{l=1}^{n}\langle AX_{l-1},\, X_l\rangle\,\langle C(X_l)y_{l-1},\, y_l\rangle\, \overline{\mathbf{E}}\left[\prod_{l=0}^{n}\langle Y_l,\, y_l\rangle \,\middle|\, \mathcal{F}_n\right]$$
$$= N^{n+1}\,\langle \pi(0),\, X_0\rangle\,\langle DX_0,\, y_0\rangle \prod_{l=1}^{n}\langle AX_{l-1},\, X_l\rangle\,\langle C(X_l)y_{l-1},\, y_l\rangle\,.$$

We can now deduce from this that

$$\overline{\mathbf{E}}\left[\overline{\Lambda}_n \mid \mathcal{F}_n\right] = N^{n+1}\,\langle \pi(0),\, X_0\rangle \prod_{l=1}^{n}\langle AX_{l-1},\, X_l\rangle\,.$$

This can be seen as follows. Set $y_n = f_1, f_2, \ldots, f_M$ in the above identity and then sum the results. Repeat this process successively for $y_{n-1}, y_{n-2}, \ldots, y_0$.

The second identity was shown in Chapter 3 in a more direct way. The lemma now follows. □

9.5 Exercises

Exercise 9.1 Show that the condition

$$\mathbf{E}\left[X_{t+1} \,|\, \mathcal{F}_t^X\right] = \mathfrak{B}_t(X_t)X_t, \qquad \text{for all } t = 0, 1, 2, \ldots,$$

implies that $\{X_t\}$ is a semi-Markov chain.

Exercise 9.2 Show that Theorem 9.3 is true without the assumption that $\tau_{ji}(m)$ does not depend on e_j.

10

Filters for Hidden Semi-Markov Models

10.1 Introduction

In this chapter estimation results for hidden semi-Markov chains are developed. These are filters and smoothers and the Viterbi algorithm. The results are then applied, using the EM algorithm, to estimate the parameters of the model. Some results in this chapter are new.

Using Koski's notation, we wish to compute

$$\hat{\pi}_i(k \mid k) = P_\theta(X_k = e_i \mid Y_{0:k} = y_{0:k}) = \frac{\alpha_i(k)}{\sum_{j=1}^{N} \alpha_j(k)}$$

where

$$\alpha_i(k) = P_\theta\left(X_k = e_i, Y_{0:k} = y_{0:k}\right).$$

It is difficult to obtain a recurrent relationship for $\alpha_i(k)$. However, we have obtained one for $\alpha_{i,d}(k)$ in Theorem 9.11:

$$\alpha_{i,d}(k) = P_\theta(X_k = e_i, h_k = d, Y_{0:k} = y_{0:k})$$

for $T_n \leq k < T_{n-1}$, $X_k = Z_n$, and $h_k = k - T_n + 1$.

Write

$$\alpha_i(k) = \sum_{d=1}^{k+1} \alpha_{i,d}(k).$$

Note that X_k is not a Markov process. However $\{(X_k, h_k), k = 0, 1, 2, \dots\}$ is a Markov process. For this latter process

$$P\left(X_{k+1} = e_j, h_{k+1} = d' \mid X_k = e_i, h_k = d \vee \mathcal{F}_{k-1}^X\right)$$

equals

$$A_{(j,d'),(i,d)} = P(X_{k+1} = e_j, h_{k+1} = d' \mid X_k = e_i, h_k = d).$$

For the case $j \neq i$, we have

$$A_{(j,d'),(i,d)} = 0$$

unless $d' = 1$ and then (see Chapter 9 for definitions)

$$A_{(j,1),(i,d)} = A_{ji}(d) \cdot \frac{p_i(d)}{F_i(d)} .$$

If $j = i$, we have

$$A_{(j,d'),(i,d)} = 0$$

unless $d' = d + 1$ and then

$$A_{(i,d+1),(i,d)} = \frac{F_i(d+1)}{F_i(d)} .$$

10.2 The Viterbi algorithm

Define

$$\delta_{k,d}(i) = \max_{x_0, x_1, \ldots, x_{k-d}} P_\theta(Y_{0:\,k} = y_{0:\,k}, X_k = e_i, h_k = d, X_{0:\,k-d} = x_{0:\,k-d})$$

and so

$$\delta_k(i) = \max_{1 \leq d \leq k+1} \delta_{k,\,d}(i) .$$

The next lemma provides the recurrence relation for $\{\delta_{k,d}(i)\}$.

Lemma 10.1 *The initial condition is given by*

$$\delta_{0,\,d}(i) = \pi_i(0)\, \langle De_i, Y_0 \rangle\, I(d = 1).$$

For $k \geq 0$ and $d \geq 2$, we have

$$\delta_{k+1,\,d}(j) = \langle C(e_j)\, y_k,\, y_{k+1} \rangle\, \frac{F_j(d)}{F_j(d-1)}\, \delta_{k,d-1}(j)$$

and for $k \geq 0$ and $d = 1$, we have

$$\delta_{k+1,\,1}(j) = \langle C(e_j) y_k, y_{k+1} \rangle \max_{i \neq j,\, 1 \leq d \leq k+1} \left\{ A_{ji}(d)\, \frac{p_i(d)}{F_i(d)}\, \delta_{k,\,d}(i) \right\} .$$

Remark In the case that X_0 does not excite Y_0, the initial condition is

$$\delta_{0,\,d}(i) = \pi_i(0)\, I(d = 1).$$

Proof Similar proofs were given in Chapter 6 and Chapter 8 for different models. For the initial condition,

$$\begin{aligned}
\delta_{0,\,d}(i) &= P_\theta\left(Y_0 = y_0,\, X_0 = e_i,\, h_0 = d\right) \\
&= \overline{\mathbf{E}}\left[M\, N\, \langle \pi(0),\, X_0\rangle \langle D X_0,\, Y_0\rangle \langle Y_0, y_0\rangle \langle X_0,\, e_i\rangle\, \mathrm{I}(h_0 = d)\right] \\
&= M\, N\, \pi_i(0)\, \langle D e_i,\, y_0\rangle\, \overline{\mathbf{E}}\left[\langle Y_0, y_0\rangle \langle X_0,\, e_i\rangle\, \mathrm{I}(h_0 = d)\right] \\
&= \pi_i(0)\, \langle D e_i,\, y_0\rangle\, \mathrm{I}(h_0 = 1)
\end{aligned}$$

as required.

We could argue more simply:

$$\begin{aligned}
\delta_{0,\,d}(i) &= P_\theta\left(Y_0 = y_0,\, X_0 = e_i,\, h_0 = d\right) \\
&= P_\theta\left(Y_0 = y_0,\, X_0 = e_i\right)\, \mathrm{I}\left(h_0 = 1\right) \\
&= P_\theta\left(Y_0 = y_0 \,|\, X_0 = e_i\right)\, P_\theta\left(X_0 = e_i\right)\, \mathrm{I}\left(h_0 = 1\right).
\end{aligned}$$

For $k \geq 0$ and $2 \leq d < k+1$ we have, using Lemma 9.1,

$$\begin{aligned}
&\delta_{k+1,\,d}(j) \\
&= \max_{x_{0:k+1-d}} P_\theta\left(Y_{0:k+1} = y_{0:k+1}, X_{k+1} = e_j, h_{k+1} = d,\right. \\
&\qquad\qquad\qquad \left. X_{0:k+1-d} = x_{0:k+1-d}\right) \\
&= \max_{x_{0:k+1-d}} \overline{\mathbf{E}}\left[\overline{\Lambda}_{k+1} \prod_{l=0}^{k+1-d} \langle X_l,\, x_l\rangle \langle X_{k+1},\, e_j\rangle \right. \\
&\qquad\qquad \left. \prod_{l=k+2-d}^{k} \langle X_l,\, e_j\rangle\, \mathrm{I}(h_{k+1} = d) \prod_{l=0}^{k+1} \langle Y_l,\, y_l\rangle \right] \\
&= \max_{x_{0:k+1-d}} \overline{\mathbf{E}}\left[\overline{\Lambda}_k\, M\, N\, \langle B_k(X_k) X_k,\, X_{k+1}\rangle \langle C(X_{k+1}) Y_k,\, Y_{k+1}\rangle \right. \\
&\left. \prod_{l=0}^{k+1-d} \langle X_l,\, x_l\rangle \langle X_{k+1},\, e_j\rangle \prod_{l=k+2-d}^{k} \langle X_l,\, e_j\rangle\, \mathrm{I}(h_{k+1} = d) \prod_{l=0}^{k+1} \langle Y_l,\, y_l\rangle \right] \\
&= \langle C(e_j)\, y_k,\, y_{k+1}\rangle \max_{x_{0:k+1-d}} \overline{\mathbf{E}}\left[\overline{\Lambda}_k \prod_{l=0}^{k} \langle Y_l,\, y_l\rangle \langle B_k(X_k) X_k,\, e_j\rangle \right. \\
&\qquad\qquad \left. \prod_{l=0}^{k-d} \langle X_l,\, x_l\rangle \prod_{l=k+1-d}^{k-1} \langle X_l,\, e_j\rangle\, \mathrm{I}(h_k = d-1)\, \langle X_k,\, e_j\rangle \right] \\
&= \langle C(e_j)\, y_k,\, y_{k+1}\rangle \frac{F_j(d)}{F_j(d-1)} \max_{x_{0:k+1-d}} \overline{\mathbf{E}}\left[\overline{\Lambda}_k \prod_{l=0}^{k} \langle Y_l,\, y_l\rangle \right. \\
&\qquad\qquad \left. \prod_{l=0}^{k-d} \langle X_l,\, x_l\rangle \prod_{l=k+1-d}^{k-1} \langle X_l,\, e_j\rangle \langle X_k,\, e_j\rangle \right]
\end{aligned}$$

$$= \langle C(e_j)\, y_k,\, y_{k+1}\rangle \, \frac{F_j(d)}{F_j(d-1)}\, \delta_{k,d-1}(j)$$

as $k+1-d = k-(d-1)$.

For $k \geq 0$ and $d = 1$ we have

$$\begin{aligned}
\delta_{k+1,\,1}(j) &= \max_{x_{0:k}} P_\theta\big(Y_{0:\,k+1} = y_{0:\,k+1}, X_{k+1} = e_j, h_{k+1} = 1, \\
&\qquad\qquad X_{0:\,k} = x_{0:\,k}\big) \\
&= \max_{x_{0:k}} \overline{\mathbf{E}}\left[\overline{\Lambda}_{k+1} \prod_{l=0}^{k+1} \langle Y_l,\, y_l\rangle\, \langle X_{k+1},\, e_j\rangle\, \mathrm{I}(h_{k+1}=1) \prod_{l=0}^{k} \langle X_l,\, x_l\rangle\right] \\
&= \max_{x_{0:k}} \overline{\mathbf{E}}\Big[\overline{\Lambda}_k\, M\, N\, \langle B_k(X_k)X_k,\, X_{k+1}\rangle\, \langle C(X_{k+1})Y_k,\, Y_{k+1}\rangle \\
&\qquad \times \prod_{l=0}^{k+1} \langle Y_l,\, y_l\rangle\, \langle X_{k+1},\, e_j\rangle\, \mathrm{I}(h_{k+1}=1) \prod_{l=0}^{k} \langle X_l,\, x_l\rangle\Big] \\
&= \langle C(e_j)\, y_k,\, y_{k+1}\rangle \\
&\qquad \times \max_{x_{0:k},\, x_k \neq e_j} \overline{\mathbf{E}}\left[\overline{\Lambda}_k\, \langle B_k(X_k)X_k,\, e_j\rangle \prod_{l=0}^{k} \langle Y_l,\, y_l\rangle \prod_{l=0}^{k} \langle X_l,\, x_l\rangle\right] \\
&= \langle C(e_j)\, y_k,\, y_{k+1}\rangle \max_{x_{0:k-d}} \max_{e_i \neq e_j\; 1 \leq d \leq k+1} \overline{\mathbf{E}}\Big[\mathrm{I}(h_k = d)\, \langle X_k,\, e_i\rangle \\
&\qquad \times \overline{\Lambda}_k\, \langle B_k(X_k)X_k,\, e_j\rangle \prod_{l=0}^{k} \langle Y_l,\, y_l\rangle \prod_{l=0}^{k} \langle X_l,\, x_l\rangle\Big] \\
&= \langle C(e_j)\, y_k,\, y_{k+1}\rangle \max_{e_i \neq e_j\; 1 \leq d \leq k+1} \Big\{ \max_{x_{0:k-d}} \overline{\mathbf{E}}\Big[\mathrm{I}(h_k = d)\, \langle X_k,\, e_i\rangle \\
&\qquad \times \overline{\Lambda}_k\, \langle B_k(X_k)X_k,\, e_j\rangle \prod_{l=0}^{k} \langle Y_l,\, y_l\rangle \prod_{l=0}^{k} \langle X_l,\, x_l\rangle\Big]\Big\} \\
&= \langle C(e_j)\, y_k,\, y_{k+1}\rangle \max_{e_i \neq e_j\; 1 \leq d \leq k+1} \Big\{ A_{j\,i}(d)\, \frac{p_j(d)}{F_j(d)} \\
&\qquad \times \max_{x_{0:k-d}} \overline{\mathbf{E}}\Big[\mathrm{I}(h_k = d)\, \langle X_k,\, e_i\rangle \overline{\Lambda}_k \prod_{l=0}^{k} \langle Y_l,\, y_l\rangle \prod_{l=0}^{k} \langle X_l,\, x_l\rangle\Big]\Big\} \\
&= \langle C(e_j)\, y_k,\, y_{k+1}\rangle \max_{e_i \neq e_j\; 1 \leq d \leq k+1} \left\{ A_{j\,i}(d)\, \frac{p_j(d)}{F_j(d)}\, \delta_{k,\,d}(i) \right\}
\end{aligned}$$

and the lemma is proved. □

Decoding

We now apply backward recursion on (i_n^*, d_n^*). Let us consider some cases:

(1) $d_n^* = 1$, then one step back we have (i_{n-1}^*, d_{n-1}^*)
(2) $d_n^* > 1$, then one step back we have $(i_{n-1}^* = i_n^*,\ d_{n-1}^* = d_n^* - 1)$
(3) $d_{n-1}^* = 1$, then one step back we have (i_{n-2}^*, d_{n-2}^*)
(4) $d_{n-1}^* > 1$, then one step back we have $(i_{n-2}^* = i_{n-1}^*, d_{n-2}^* = d_{n-1}^* - 1)$

and so on.

10.3 Smoothers

We wish to compute for each $1 \le i \le N$,

$$\hat{\pi}_i(k|\, n) = P_\theta(X_k = e_i|\, Y_{0:\, n} = y_{0:\, n})$$

for $0 \le k \le n$. This is Koski's notation. An alternative notation is to use $\gamma_i(k|n)$ which seems tidier.

Once the smoothers have been computed (estimated) for a fixed number n of observations, for each k, we can choose i_k^* so that

$$\hat{\pi}_{i_k^*}(k|\, n) = \max_{1 \le i \le N} P_\theta(X_k = e_i|\, Y_{0:\, n} = y_{0:\, n})\,.$$

This is called the Maximal-a-Posteriori (MAP) method of decoding.

The problem with this approach is that $(i_0^*, i_1^*, \ldots, i_n^*)$ may not be feasible. However, if $A_{ji}(d) > 0$ for all $j \neq i$ and $d \ge 1$, then the MAP and Viterbi methods should provide the same decoding.

We shall see later that smoothers are also used in the EM algorithm. For example, $\hat{\pi}_i(0) = \hat{\pi}_i(0|n)$.

As with filters, we define

$$\alpha_{k,\, d}(i) = P_\theta(X_k = e_i, h_k = d, Y_{0:\, n} = y_{0:\, n})$$

and

$$\beta_{k,\, d}(i) = P_\theta(Y_{k+1:\, n} = y_{k+1:\, n}|\, X_k = e_i, h_k = d)$$

where in the latter we need to assume that $k + 1 \le n$.

Backward recursion for β

Lemma 10.2 (1) *For all d and j, we have $\beta_{n,\,d}(j) = 1$.*
(2) *We have*

$$\beta_{k,\,d}(j) = \sum_{r \neq j} \langle C(e_r)\, y_k, y_{k+1} \rangle \cdot A_{rj}(d) \cdot \frac{p_j(d)}{F_j(d)} \cdot \beta_{k+1,\,1}(r) + \langle C(e_j)\, y_k, y_{k+1} \rangle \cdot \frac{F_j(d+1)}{F_j(d)} \cdot \beta_{k+1,\,d+1}(j)\,.$$

We first prove the following lemma.

Lemma 10.3 *For $k+1 \leq n$, $1 \leq d \leq k+1$ and $1 \leq j \leq N$,*

$$\beta_{k,\,d}(j) = M^{-n+k}\, \overline{\mathbf{E}}\left[\overline{\Lambda}_{k+1:n} \mid Y_{0:n} = y_{0:n},\, X_k = e_j,\, h_k = d\right].$$

Proof Using the conditional Bayes identity, the left-hand side is (where for readability we have suppressed limits $l = k+1$ to n on the product, and denoted $X_k = e_j,\ h_k = d$ by Ψ)

$$\begin{aligned}
&\mathbf{E}_\theta\Big[\prod \langle Y_l, y_l \rangle \Big| \Psi\Big] \\
&= \frac{\overline{\mathbf{E}}\left[\overline{\Lambda}_n \prod \langle Y_l,\, y_l \rangle \middle|\, \Psi\right]}{\overline{\mathbf{E}}\left[\overline{\Lambda}_n \middle|\, \Psi\right]} \\
&= \frac{\overline{\mathbf{E}}\left[\overline{\Lambda}_n \prod \langle Y_l,\, y_l \rangle \,\middle|\, \Psi\right]}{\overline{\mathbf{E}}\left[\overline{\Lambda}_k \,\middle|\, \Psi\right]} \\
&= \frac{\overline{\mathbf{E}}\left[\overline{\mathbf{E}}\left[\overline{\Lambda}_n \prod \langle Y_l,\, y_l \rangle \,\middle|\, \mathcal{Y}_n \vee \mathcal{F}_k\right] \middle|\, \Psi\right]}{\overline{\mathbf{E}}\left[\overline{\Lambda}_k \,\middle|\, \Psi\right]} \\
&= \frac{\overline{\mathbf{E}}\Big[\overline{\Lambda}_k \prod \langle Y_l,\, y_l \rangle\, \overline{\mathbf{E}}\Big[\overline{\Lambda}_{k+1:n} \,\Big|\, \mathcal{Y}_n \vee X_k \vee h_k\Big] \,\Big|\, \Psi\Big]}{\overline{\mathbf{E}}\Big[\overline{\Lambda}_k \,\Big|\, \Psi\Big]} \\
&= \frac{\overline{\mathbf{E}}\left[\overline{\Lambda}_k \prod \langle Y_l,\, y_l \rangle\, \overline{\mathbf{E}}\left[\overline{\Lambda}_{k+1:n} \,\middle|\, Y_{k+1:n} \vee X_k \vee h_k\right] \middle|\, \Psi\right]}{\overline{\mathbf{E}}\left[\overline{\Lambda}_k \,\middle|\, \Psi\right]} \\
&= \frac{\overline{\mathbf{E}}\left[\overline{\Lambda}_k \prod \langle Y_l, y_l \rangle \middle|\, \Psi\right] \overline{\mathbf{E}}\left[\overline{\Lambda}_{k+1:n} \middle|\, Y_{k+1:n} = y_{k+1:n}, \Psi\right]}{\overline{\mathbf{E}}\left[\overline{\Lambda}_k \middle|\, \Psi\right]} \\
&= \frac{M^{-n+k}\overline{\mathbf{E}}\left[\overline{\Lambda}_k \middle|\, X_k = e_j, h_k = d\right] \overline{\mathbf{E}}\left[\overline{\Lambda}_{k+1:n} \middle|\, Y_{k+1:n} = y_{k+1:n},\, \Psi\right]}{\overline{\mathbf{E}}\left[\overline{\Lambda}_k \middle|\, \Psi\right]} \\
&= M^{-n+k}\, \overline{\mathbf{E}}\left[\overline{\Lambda}_{k+1:n} \,\middle|\, Y_{k+1:n} = y_{k+1:n}, \Psi\right] \\
&= M^{-n+k}\, \overline{\mathbf{E}}\left[\overline{\Lambda}_{k+1:n} \,\middle|\, Y_{0:n} = y_{0:n}, \Psi\right]
\end{aligned}$$

and the lemma is proved. □

Remark It is convenient to set

$$v^j_{k,n}(y_{0:n}) = \overline{\mathbf{E}}\left[\overline{\Lambda}_{k+1:n} \mid Y_{0:n} = y_{0:n},\ X_k = e_j,\ h_k = d\right].$$

It is easier to find a recurrence relation for this quantity.

We now revert to the proof of Lemma 10.2.

Proof of Lemma 10.2. We shall first prove the recurrence relationship. If $k < n-1$, then

$$\begin{aligned}
&\beta_{k,\,d}(j)\\
&= M^{-n+k}\,\overline{\mathbf{E}}\left[\overline{\Lambda}_{k+1:n} \mid Y_{0:n} = y_{0:n},\ X_k = e_j,\ h_k = d\right]\\
&= M^{-n+k}\,\overline{\mathbf{E}}\big[\overline{\Lambda}_{k+2:n}\, M\, N\, \langle B_k(X_k)\, X_k,\ x_{k+1}\rangle \times\\
&\qquad\qquad \times\ \langle C(X_{k+1})\, Y_k,\ Y_{k+1}\rangle \mid \text{ditto}\big]\\
&= M^{-n+k}\sum_{r=1}^{N}\overline{\mathbf{E}}\big[\overline{\Lambda}_{k+2:n}\langle X_{k+1}, e_r\rangle\, MN\langle B_k(X_k)X_k, X_{k+1}\rangle \times\\
&\qquad\qquad \times\ \langle C(X_{k+1})Y_k, Y_{k+1}\rangle \mid \text{ditto}\big]\\
&= M^{-n+k+1}N\sum_{r=1}^{N}\langle C(e_r)y_k, y_{k+1}\rangle\overline{\mathbf{E}}\big[\overline{\Lambda}_{k+2:n}\langle X_{k+1}, e_r\rangle \times\\
&\qquad\qquad \times\ \langle B_k(X_k)X_k, X_{k+1}\rangle \mid \text{ditto}\big]\\
&= M^{-n+k+1}N\sum_{r=1, r\neq j}^{N} A_{rj}(d)\frac{p_j(d)}{F_j(d)}\langle C(e_r)y_k, y_{k+1}\rangle \times\\
&\qquad\qquad \times\ \overline{\mathbf{E}}\left[\overline{\Lambda}_{k+2:n}\langle X_{k+1}, e_r\rangle \mid \text{ditto}\right]\\
&\quad + M^{-n+k+1}\, N\, \frac{F_j(d+1)}{F_j(d)}\langle C(e_j)\, y_k,\ y_{k+1}\rangle \times\\
&\qquad\qquad \times\ \overline{\mathbf{E}}\left[\overline{\Lambda}_{k+2:n}\langle X_{k+1}, e_j\rangle \mid \text{ditto}\right]\\
&\equiv \mathrm{T}_1 + \mathrm{T}_2.
\end{aligned}$$

For T_1, we use for $r \neq j$,

$$\begin{aligned}
&\overline{\mathbf{E}}\left[\overline{\Lambda}_{k+2:n}\,\langle X_{k+1},\ e_r\rangle \mid \text{ditto}\right]\\
&\quad = \overline{\mathbf{E}}\left[\overline{\mathbf{E}}\left[\overline{\Lambda}_{k+2:n}\,\langle X_{k+1},\ e_r\rangle \mid \mathcal{Y}_n \vee \mathcal{F}_{k+1}\right] \mid \text{ditto}\right]\\
&\quad = \overline{\mathbf{E}}\left[\langle X_{k+1},\ e_r\rangle\,\overline{\mathbf{E}}\left[\overline{\Lambda}_{k+2:n} \mid \mathcal{Y}_n,\ X_{k+1} = e_r,\ h_{k+1} = 1\right] \mid \text{ditto}\right]\\
&\quad = M^{n-1-k}\,\beta_{k+1,1}(r)\,\overline{\mathbf{E}}\left[\langle X_{k+1},\ e_r\rangle \mid \text{ditto}\right]\\
&\quad = M^{n-1-k}\, N^{-1}\,\beta_{k+1,1}(r).
\end{aligned}$$

For T_2,

$$\begin{aligned}
&\overline{\mathbf{E}}\left[\overline{\Lambda}_{k+2:n}\,\langle X_{k+1},\,e_j\rangle\,\middle|\,\text{ditto}\right]\\
&=\overline{\mathbf{E}}\left[\overline{\mathbf{E}}\left[\overline{\Lambda}_{k+2:n}\,\langle X_{k+1},\,e_j\rangle\,\middle|\,\mathcal{Y}_n\vee\mathcal{F}_{k+1}\right]\middle|\,\text{ditto}\right]\\
&=\overline{\mathbf{E}}\left[\langle X_{k+1},e_j\rangle\overline{\mathbf{E}}\left[\overline{\Lambda}_{k+2:n}\middle|\,\mathcal{Y}_n\vee\{X_{k+1}=e_r\}\vee\{h_{k+1}=1\}\right]\middle|\,\text{ditto}\right]\\
&=\overline{\mathbf{E}}\left[\overline{\Lambda}_{k+2:n}\middle|\,\mathcal{Y}_n\vee\{X_{k+1}=e_r\}\vee\{h_{k+1}=1\}\right]\overline{\mathbf{E}}\left[\langle X_{k+1},e_j\rangle\middle|\,\text{ditto}\right]\\
&=M^{n-1-k}\,N^{-1}\beta_{k+1,d+1}(j).
\end{aligned}$$

Combining these results we obtain for $k+1\leq n-1$,

$$\begin{aligned}
\beta_{k,d}(j)&=\mathrm{T}_1+\mathrm{T}_2\\
&=\sum_{r=1,\,r\neq j}^{N}A_{r\,j}(d)\,\frac{p_j(d)}{F_j(d)}\langle C(e_r)\,y_k,\,y_{k+1}\rangle\,\beta_{k+1,1}(r)\\
&\quad+\frac{F_j(d+1)}{F_j(d)}\,\langle C(e_j)\,y_k,\,y_{k+1}\rangle\,\beta_{k+1,d+1}(j).
\end{aligned}$$

We have

$$\begin{aligned}
&\beta_{n-1,d}(j)\\
&=M^{-1}\,\overline{\mathbf{E}}\left[\overline{\Lambda}_{n:n}\,|\,Y_{0:n}=y_{0:n},\,X_{n-1}=e_j,\,h_{n-1}=d\right]\\
&=M^{-1}\overline{\mathbf{E}}\left[M\,N\,\langle B_{n-1}(X_{n-1})\,X_{n-1},\,X_n\rangle\,\langle C(X_n)\,Y_{n-1},\,Y_n\rangle\,|\,\text{ditto}\right]\\
&=N\sum_{r=1}^{N}\overline{\mathbf{E}}\left[\langle X_n,e_r\rangle\langle B_{n-1}(X_{n-1})X_{n-1},X_n\rangle\langle C(X_n)Y_{n-1},Y_n\rangle\,|\,\text{ditto}\right]\\
&=N\sum_{r=1}^{N}\langle C(e_r)y_{n-1},y_n\rangle\overline{\mathbf{E}}\left[\langle X_n,e_r\rangle\langle B_{n-1}(X_{n-1})X_{n-1},X_n\rangle\,|\,\text{ditto}\right]\\
&=N\sum_{r=1,r\neq j}^{N}\langle C(e_r)y_{n-1},y_n\rangle\overline{\mathbf{E}}\left[\langle X_n,e_r\rangle\langle B_{n-1}(X_{n-1})X_{n-1},X_n\rangle\,|\,\text{ditto}\right]\\
&\quad+N\,\langle C(e_j)\,y_{n-1},\,y_n\rangle\,\overline{\mathbf{E}}\left[\langle X_n,\,e_j\rangle\,\langle B_{n-1}(X_{n-1})\,X_{n-1},\,X_n\rangle\,|\,\text{ditto}\right]\\
&=N\sum_{r=1,r\neq j}^{N}\langle C(e_r)y_{n-1},y_n\rangle A_{nj}(d)\frac{p_j(d)}{F_j(d)}\overline{\mathbf{E}}\left[\langle X_n,e_r\rangle\,|\,\text{ditto}\right]\\
&\quad+N\,\langle C(e_j)\,y_{n-1},\,y_n\rangle\,\frac{F_j(d+1)}{F_j(d)}\,\overline{\mathbf{E}}\left[\langle X_n,\,e_j\rangle\,|\,\text{ditto}\right]\\
&=\sum_{r=1,r\neq j}^{N}\langle C(e_r)y_{n-1},y_n\rangle A_{nj}(d)\frac{p_j(d)}{F_j(d)}+\langle C(e_j)\,y_{n-1},y_n\rangle\frac{F_j(d+1)}{F_j(d)}
\end{aligned}$$

and this implies that the backward recursion holds with $\beta_{n,d}(j) = 1$ for all $1 \leq j \leq N$ and $1 \leq d \leq n+1$. The lemma is proved. □

Remark The recurrence for β is backward. If a new observation is made at $n+1$, it means the backward recursion would normally be made starting at $n+1$. As explained in Chapter 8, this can be circumvented as follows:

$$\beta_l = \Gamma\, B(y_{l+1},\, y_l)\, \beta_{l+1} \tag{10.1}$$

where

$$\beta_l = \{\beta_{l,d}(j)\}$$
$$B(y_{l+1},\, y_l) = \operatorname{diag}\left(\langle C(e_j)\, y_l,\, y_{l+1}\rangle\right)$$

and

$$\Gamma_{(j,d),\,(r,d')} = \begin{cases} A_{rj}\frac{p_j(d)}{F_j(d)}\, \mathrm{I}(d'=1) & \text{if } r \neq j \\ \frac{F_j(d+1)}{F_j(d)}\, \mathrm{I}(d'=d+1) & \text{if } r = j \end{cases}$$

where (10.1) can be interpreted in $\mathbb{R}^N$

$$\beta_{l,d}(j) = \sum_{r,d'} \Gamma_{(j,d),\,(r,d')}\, B(y_{l+1},\, y_l)_r\, \beta_{l+1,d'}(r).$$

We now define

$$\Phi_{k,n} = \Gamma\, B(y_{k+1},\, y_k)\, \Gamma\, B(y_{k+2},\, y_{k+1}) \cdots \Gamma\, B(y_n,\, y_{n-1})$$

for which

$$\Phi_{k,n+1} = \Phi_{k,n}\, \Gamma\, B(y_{n+1},\, y_n)$$

and then

$$\beta_k = \Phi_{k,n+1}\, \mathbf{1}$$

where $\mathbf{1} = 1$ for all (j,d) gives β_k bases on $n+1$ observations.

The forward recursion for d was discussed under filtering.

Lemma 10.4 (Baum's Identity)

$$P_\theta\left(Y_{0:n} = y_{0:n},\, X_k = e_j,\, h_k = d\,\right) = \alpha_{k,d}(j)\, \beta_{k,d}(j).$$

Proof We have

$$\begin{aligned} &P_\theta\left(Y_{0:n} = y_{0:n},\, X_k = e_j,\, h_k = d\,\right) \\ &\quad = \mathbf{E}_\theta\left[\prod_{l=0}^{n} \langle Y_l,\, y_l\rangle\, \langle X_k,\, e_j\rangle\, \mathrm{I}(h_k = d)\right] \end{aligned}$$

$$
\begin{aligned}
&= \mathbf{E}_\theta \left[\mathbf{E}_\theta \left[\prod_{l=0}^{n} \langle Y_l,\, y_l\rangle \, \langle X_k,\, e_j\rangle \, \mathrm{I}(h_k = d) \,\middle|\, \mathcal{Y}_k \vee \mathcal{F}_k \right]\right] \\
&= \mathbf{E}_\theta \left[\prod_{l=0}^{k} \langle Y_l,\, y_l\rangle \, \langle X_k,\, e_j\rangle \, \mathrm{I}(h_k = d) \, \mathbf{E}_\theta \left[\prod_{l=k+1}^{n} \langle Y_l,\, y_l\rangle \,\middle|\, \mathcal{Y}_k \vee \mathcal{F}_k \right]\right] \\
&= \mathbf{E}_\theta \left[\prod_{l=0}^{k} \langle Y_l,\, y_l\rangle \, \langle X_k,\, e_j\rangle \, \mathrm{I}(h_k = d) \, \mathbf{E}_\theta \left[\prod_{l=k+1}^{n} \langle Y_l,\, y_l\rangle \,\middle|\, \mathcal{F}_k \right]\right] \\
&= \mathbf{E}_\theta \left[\prod_{l=0}^{k} \langle Y_l,\, y_l\rangle \, \langle X_k,\, e_j\rangle \, \mathrm{I}(h_k = d) \, \mathbf{E}_\theta \left[\prod_{l=k+1}^{n} \langle Y_l,\, y_l\rangle \,\middle|\, X_k,\, h_k \right]\right] \\
&= \mathbf{E}_\theta \left[\prod_{l=0}^{k} \langle Y_l,\, y_l\rangle \, \langle X_k,\, e_j\rangle \, \mathrm{I}(h_k = d) \right] \times \\
&\qquad\qquad \times \; \mathbf{E}_\theta \left[\prod_{l=k+1}^{n} \langle Y_l,\, y_l\rangle \,\middle|\, X_k = e_j,\, h_k = d \right] \\
&= \alpha_{k,d}(j)\, \beta_{k,d}(j)
\end{aligned}
$$

and the identity is proved. □

Corollary 10.5 *For each $n = 0, 1, 2, \ldots$*

$$P_\theta\left(Y_{0:n} = y_{0:n}\right) = \sum_{j=1}^{N} \sum_{d=1}^{n+1} \alpha_{k,d}(j)\, \beta_{k,d}(j).$$

Remark We note the right-hand side of this expression does not depend on k.

Corollary 10.6 *For each $n = 0, 1, 2, \ldots$ and $0 \le k \le n$,*

$$P_\theta\left(X_k = e_j,\, h_k = d \,|\, Y_{0:n} = y_{0:n}\right) = \frac{\alpha_{k,d}(j)\, \beta_{k,d}(j)}{\sum_{j=1}^{N} \sum_{d=1}^{n+1} \alpha_{k,d}(j)\, \beta_{k,d}(j)}$$

for $1 \le j \le N$ and $1 \le d \le n+1$. □

Corollary 10.7 *For each $n = 0, 1, 2, \ldots$ and $0 \le k \le n$,*

$$
\begin{aligned}
\hat{\pi}_j(k|\, n) &= \sum_{d=1}^{n+1} P_\theta\left(X_k = e_j,\, h_k = d \,|\, Y_{0:n} = y_{0:n}\right) \\
&= \frac{\sum_{d=1}^{n+1} \alpha_{k,d}(j)\, \beta_{k,d}(j)}{\sum_{j=1}^{N} \sum_{d=1}^{n+1} \alpha_{k,d}(j)\, \beta_{k,d}(j)}
\end{aligned}
$$

for $1 \le j \le N$. □

Corollary 10.7 gives the usual smoother formula based on backward and forward recursions. With the notation

$$\hat{\pi}_{j,d}(k|\, n) = P_\theta\left(X_k = e_j,\, h_k = d \,|\, Y_{0:n} = y_{0:n}\right)$$

we have the following Derin-type formula. In the proof of the next lemma we use an extension of Lemma 9.12.

Lemma 10.8 *For any $n \geq 1$ and $y_l \in \{f_1, f_2, \ldots, f_M\}$ for $0 \leq l \leq n$,*

$$P_\theta(Y_{0:n} = y_{0:n} \,|\, \mathcal{F}_n) = \langle D\, X_0,\, y_0\rangle \prod_{l=1}^{n} \langle C(X_l)\, y_{l-1},\, y_l\rangle$$

for the hidden SM–M1 *model, and for the hidden* SM–M0 *model, we have*

$$P_\theta(Y_{0:n} = y_{0:n} \,|\, \mathcal{F}_n) = \prod_{l=0}^{n} \langle C\, X_l,\, y_l\rangle\, .$$

Proof The proof is almost the same as the proof of Lemma 9.12 which was given for the hidden Markov chain models introduced in Chapter 8.

We shall just indicate the changes needed in the proof for the SM–M1 model. Corresponding changes apply to the SM–M0 case.

We have for the SM–M1 model

$$\overline{\mathbf{E}}\left[\overline{\Lambda}_n \prod_{l=0}^{n} \langle Y_l,\, y_l\rangle \,\middle|\, \mathcal{F}_n\right]$$

$$= N^{n+1} \langle \pi(0),\, X_0\rangle \langle D X_0, y_0\rangle \prod_{l=1}^{n} \langle B_{l-1}(X_{l-1})\, X_{l-1}, X_l\rangle \langle C(X_l) y_{l-1}, y_l\rangle\, .$$

and

$$\overline{\mathbf{E}}\left[\overline{\Lambda}_n \,|\, \mathcal{F}_n\right] = N^{n+1}\, \langle \pi(0),\, X_0\rangle \prod_{l=1}^{n} \langle B_{l-1}(X_{l-1})\, X_{l-1},\, X_l\rangle\, .$$

from which the lemma follows. □

Lemma 10.9 (Derin's Formula) *For $n = 0, 1, 2, \ldots$, $0 \leq k \leq n-1$, $1 \leq j \leq N$ and $1 \leq d \leq n+1$,*

$$\hat{\pi}_{j,d}(k|\, n) = \hat{\pi}_{j,d}(k|\, k) \sum_{i=1}^{N} \sum_{d'=1}^{n+1} A_{(i,d'),(j,d)}\, \frac{\hat{\pi}_{i,d'}(k+1|\, n)}{\hat{\pi}_{i,d'}(k+1|\, k)}$$

where in this sum, if $d' = 1$ then $i \neq j$ and if $d' > 1$ then $i = j$ and $d' = d - 1$.

Proof We present the proof for the HSMM–1 case. The proof of the HSMM–0 case is similar.

We have

$$\begin{aligned}
&\hat{\pi}_{j,d}(k|\, n)\\
&= P_\theta\left(X_k = e_j,\, h_k = d \,|\, Y_{0:n} = y_{0:n}\right)\\
&= \sum_{i,d'} P_\theta\left(X_k = e_j,\, h_k = d,\, X_{k+1} = e_i,\, h_{k+1} = d' \,|\, Y_{0:n} = y_{0:n}\right)\\
&= \sum_{i,d'} P_\theta\left(X_k = e_j,\, h_k = d,\, |\, X_{k+1} = e_i,\, h_{k+1} = d',\, Y_{0:n} = y_{0:n}\right)\\
&\qquad\qquad \times P_\theta\left(X_{k+1} = e_i,\, h_{k+1} = d' \,|\, Y_{0:n} = y_{0:n}\right)\\
&= \sum_{i,d'} P_\theta\left(X_k = e_j,\, h_k = d,\, |\, X_{k+1} = e_i,\, h_{k+1} = d',\, Y_{0:n} = y_{0:n}\right) \times\\
&\qquad\qquad \times\ \hat{\pi}_{i,d'}(k+1|\, n)
\end{aligned}$$

From now on, we can assume to deal only with terms in this sum for which

$$P_\theta\left(X_{k+1} = e_i,\, h_{k+1} = d',\, Y_{0:n} = y_{0:n}\right) \neq 0.$$

Write

$$P_\theta\left(X_k = e_j,\, h_k = d,\, |\, X_{k+1} = e_i,\, h_{k+1} = d',\, Y_{0:n} = y_{0:n}\right)$$

as

$$\frac{P_\theta\left(X_k = e_j,\, h_k = d,\, X_{k+1} = e_i,\, h_{k+1} = d',\, Y_{0:n} = y_{0:n}\right)}{P_\theta\left(X_{k+1} = e_i,\, h_{k+1} = d',\, Y_{0:n} = y_{0:n}\right)}. \tag{10.2}$$

We now compute the numerator of this expression.

Note

$$P_\theta\left(X_k = e_j,\, h_k = d,\, X_{k+1} = e_i,\, h_{k+1} = d',\, Y_{0:n} = y_{0:n}\right)$$

equals the product of

$$P_\theta\left(Y_{0:n} = y_{0:n} \,|\, X_k = e_j,\, h_k = d,\, X_{k+1} = e_i,\, h_{k+1} = d'\right)$$

and

$$P_\theta\left(X_k = e_j,\, h_k = d,\, X_{k+1} = e_i,\, h_{k+1} = d'\right)$$

The second term is

$$A_{(i,d'),(j,d)}\, P_\theta\left(X_k = e_j,\, h_k = d\right)$$

The first term

$$P_\theta\left(Y_{0:n} = y_{0:n} \,|\, X_k = e_j,\, h_k = d,\, X_{k+1} = e_i,\, h_{k+1} = d'\right)$$

$$
\begin{aligned}
&= \mathbf{E}_\theta\left[\prod_{l=0}^{n}\langle Y_l,\, y_l\rangle \,\middle|\, \text{ditto}\right]\\
&= \mathbf{E}_\theta\left[\mathbf{E}_\theta\left[\prod_{l=0}^{n}\langle Y_l,\, y_l\rangle \,\middle|\, \mathcal{F}_n\right] \,\middle|\, \text{ditto}\right]\\
&= \mathbf{E}_\theta\left[\prod_{l=1}^{n}\langle C(X_l)\, y_{l-1},\, y_l\rangle\langle D\, X_0,\, y_0\rangle \,\middle|\, \text{ditto}\right]\\
&= \mathbf{E}_\theta\left[\mathbf{E}_\theta\left[\prod_{l=1}^{n}\langle C(X_l)\, y_{l-1},\, y_l\rangle\langle D\, X_0,\, y_0\rangle \,\middle|\, \mathcal{F}_{k+1}\right] \,\middle|\, \text{ditto}\right]\\
&= \mathbf{E}_\theta\Bigg[\prod_{l=1}^{k}\langle C(X_l)\, y_{l-1},\, y_l\rangle\langle D\, X_0,\, y_0\rangle \times\\
&\qquad \times\ \mathbf{E}_\theta\left[\prod_{l=k+1}^{n}\langle C(X_l)\, y_{l-1},\, y_l\rangle \,\middle|\, \mathcal{F}_{k+1}\right] \,\Bigg|\, \text{ditto}\Bigg]\\
&= \mathbf{E}_\theta\left[\prod_{l=k+1}^{n}\langle C(X_l)\, y_{l-1},\, y_l\rangle \,\middle|\, X_{k+1} = e_i,\, h_{k+1} = d'\right]\\
&\quad \times \mathbf{E}_\theta\left[\prod_{l=1}^{k}\langle C(X_l)\, y_{l-1},\, y_l\rangle\langle D\, X_0,\, y_0\rangle \,\middle|\, X_k = e_j,\, h_k = d\right]\\
&= \mathbf{E}_\theta\left[\prod_{l=k+1}^{n}\langle C(X_l)\, y_{l-1},\, y_l\rangle \,\middle|\, X_{k+1} = e_i,\, h_{k+1} = d'\right]\\
&\qquad \times \mathbf{E}_\theta\left[\prod_{l=0}^{k}\langle Y_l,\, y_l\rangle, \,\middle|\, X_k = e_j,\, h_k = d\right]\\
&= \mathbf{E}_\theta\left[\prod_{l=k+1}^{n}\langle C(X_l)\, y_{l-1},\, y_l\rangle \,\middle|\, X_{k+1} = e_i,\, h_{k+1} = d'\right]\\
&\qquad \times P_\theta\left(Y_{0:k} = y_{0:k} \,|\, X_k = e_j,\, h_k = d\right)
\end{aligned}
$$

where, of course we have twice used a version of Lemma 10.8.

So the numerator of (10.2) is given by

$$
\begin{aligned}
&A_{(i,d'),(j,d)}\, \hat{\pi}_{j,d}(k|k)\, P_\theta(Y_{0:k} = y_{0:k}) \times\\
&\quad \times\ \mathbf{E}_\theta\left[\prod_{l=k+1}^{n}\langle C(X_l)\, y_{l-1},\, y_l\rangle \,\middle|\, X_{k+1} = e_i,\, h_{k+1} = d'\right]
\end{aligned}
$$

The denominator of (10.2) is given by

$$
P_\theta\left(Y_{0:n} = y_{0:n} \,|\, X_{k+1} = e_i,\, h_{k+1} = d'\right)\, P_\theta\left(X_{k+1} = e_i,\, h_{k+1} = d'\right)
$$

and the first term is the same as

$$P_\theta\left(Y_{0:k} = y_{0:k} \mid X_{k+1} = e_i,\, h_{k+1} = d'\right) \times \\ \times\ \mathbf{E}_\theta\left[\prod_{l=k+1}^{n} \langle C(X_l)\, y_{l-1},\, y_l\rangle \,\middle|\, X_{k+1} = e_i,\, h_{k+1} = d'\right].$$

Therefore, the denominator of (10.2) is

$$\hat{\pi}_{i,d'}(k+1|k)\; P_\theta\left(Y_{0:k} = y_{0:k}\right) \times \\ \times\ \mathbf{E}_\theta\left[\prod_{l=k+1}^{n} \langle C(X_l)\, y_{l-1},\, y_l\rangle \,\middle|\, X_{k+1} = e_i,\, h_{k+1} = d'\right]$$

and so the expression (10.2) equals:

$$A_{(i,d'),(j,d)}\, \frac{\hat{\pi}_{j,d}(k|k)}{\hat{\pi}_{i,d'}(k+1|k)}$$

and the lemma is proved. □

Remark 10.10 In order to apply this recursion, we need an expression for $\hat{\pi}_{i,d'}(k+1|k)$. This is provided by

$$\hat{\pi}_{i,d'}(k+1|k) = \sum_{l,r} A_{(i,d'),(l,r)}\, \hat{\pi}_{l,r}(k|k)$$

This can be established as follows:

$$\begin{aligned} &\hat{\pi}_{i,d'}(k+1|k) \\ &\quad = P_\theta\left(X_{k+1} = e_i, h_{k+1} = d' \mid Y_{0:k} = y_{0:k}\right) \\ &\quad = \sum_{l,r} P_\theta\left(X_{k+1} = e_i, h_{k+1} = d',\, X_k = e_l,\, h_k = r \mid Y_{0:k} = y_{0:k}\right) \\ &\quad = \sum_{l,r} \frac{P_\theta\left(X_{k+1} = e_i, h_{k+1} = d',\, X_k = e_l,\, h_k = r,\, Y_{0:k} = y_{0:k}\right)}{P_\theta\left(Y_{0:k} = y_{0:k}\right)} \end{aligned}$$

and

$$\begin{aligned} &P_\theta\left(X_{k+1} = e_i, h_{k+1} = d',\, X_k = e_l,\, h_k = r,\, Y_{0:k} = y_{0:k}\right) \\ &\quad = P_\theta\left(X_{k+1} = e_i, h_{k+1} = d' \mid X_k = e_l,\, h_k = r,\, Y_{0:k} = y_{0:k}\right) \times \\ &\qquad \times\ P_\theta\left(X_k = e_l,\, h_k = r,\, Y_{0:k} = y_{0:k}\right) \\ &\quad = P_\theta\left(X_{k+1} = e_i, h_{k+1} = d' \,|X_k = e_l, h_k = r\right) \times \\ &\qquad \times\ P_\theta\left(X_k = e_l, h_k = r, Y_{0:k} = y_{0:k}\right) \end{aligned}$$

and so the claim holds.

Remark 10.11 Note that

$$P_\theta(X_k = e_j, h_k = d \,|\, Y_{0:n} = y_{0:n}) = \frac{P_\theta(X_k = e_j, h_k = d, Y_{0:n} = y_{0:n})}{L_\theta}$$

where the likelihood is given by

$$L_\theta = P_\theta(Y_{0:n} = y_{0:n}) = \sum_{r=1}^{N} \sum_{d=1}^{k+1} P_\theta(X_k = e_j, h_k = d, Y_{0:n} = y_{0:n})\,.$$

We can apply

$$P_\theta(X_k = e_j, h_k = d, Y_{0:n} = y_{0:n}) = \alpha_{k,\,d}(j)\, \beta_{k,\,d}(j)$$

for both SM–M0 and SM–M1. However in SM–M0, $\langle C(e_i)\, y_k, y_{k+1}\rangle$ is replaced with $\langle C(e_i), y_{k+1}\rangle$.

10.4 EM algorithm for estimating a hidden semi-Markov model

We need to estimate the parameters $p_j(d), A_{ji}(d), C, \pi(0)$, and D. To estimate the initial emission D, we need multi-observations; see the remarks in Chapter 6. Given a current estimate of parameter θ, we seek θ' to maximize

$$\mathbf{E}_\theta \left[\log \bar{\Lambda}_n(\theta') |\, Y_{0:n} = y_{0:n}\right]$$

and we hope that $L_{\theta'} > L_\theta$.

We shall refer to θ' as an updated value of the the parameter when computed in this way.

The theory behind the EM algorithm was given in Chapter 6. We shall not discuss some of the theoretical issues: do the iterations from the EM algorithm converge? If they do, then the limit would seem to optimize the conditional log-likelihood, but perhaps not the likelihood in general. There is also the theoretical issue: should we be maximizing the likelihood as as estimation procedure or should we be optimizing the conditional log-likelihood. The latter is related to as the information criteria for model estimation.

Recall that

$$\log \bar{\Lambda}_n(\theta') = \text{constant} + \log \bar{\Lambda}_n^{\pi}(\theta') + \log \bar{\Lambda}_n^{D}(\theta') + \log \bar{\Lambda}_n^{C}(\theta') + \log \bar{\Lambda}_n^{A}(\theta')$$

where the constant does not depend on θ' and

- $\bar{\Lambda}_n^{\pi}(\theta') = \langle \pi'(0), X_0\rangle$

- $\bar{\Lambda}_n^D(\theta') = \langle D' X_0, y_0 \rangle$
- $\bar{\Lambda}_n^C(\theta') = \prod_{l=1}^{n} \langle C'(X_l) y_{l-1}, y_l \rangle$
- $\bar{\Lambda}_n^A(\theta') = \prod_{l=1}^{n} \langle B'_{l-1}(X_{l-1}) X_{l-1}, X_l \rangle$.

We now seek $\theta' = (\pi', D', C', A')$ to maximize respectively

$$\begin{aligned}
J_1(\pi') &= \mathbf{E}_\theta \left[\log\langle \pi'(0), X_0 \rangle \mid Y_{0:n} = y_{0:n} \right] \\
J_2(D') &= \mathbf{E}_\theta \left[\log\langle D' X_0, y_0 \rangle \mid Y_{0:n} = y_{0:n} \right] \\
J_3(C') &= \mathbf{E}_\theta \left[\log \prod_{l=1}^{n} \langle C'(X_l) y_{l-1}, y_l \rangle \,\middle|\, Y_{0:n} = y_{0:n} \right] \\
J_4(A') &= \mathbf{E}_\theta \left[\log \prod_{l=1}^{n} \langle B'_{l-1}(X_{l-1}) X_{l-1}, X_l \rangle \,\middle|\, Y_{0:n} = y_{0:n} \right].
\end{aligned}$$

We now treat these optimization problems in turn.

Note that

$$\begin{aligned}
J_1(\pi') &= \mathbf{E}_\theta \left[\log\langle \pi'(0), X_0 \rangle \mid Y_{0:n} = y_{0:n} \right] \\
&= \sum_{i=1}^{N} \mathbf{E}_\theta \left[\langle X_0,\, e_i \rangle \, \log\langle \pi'(0), X_0 \rangle \mid Y_{0:n} = y_{0:n} \right] \\
&= \sum_{i=1}^{N} \log \pi'_i(0) \, \mathbf{E}_\theta \left[\langle X_0,\, e_i \rangle \mid Y_{0:n} = y_{0:n} \right] \\
&= \sum_{i=1}^{N} \log \pi'_i(0) \, P_\theta \left(X_0 = e_i \mid Y_{0:n} = y_{0:n} \right) \\
&= \sum_{i=1}^{N} \log \pi'_i(0) \, \hat{\pi}_i(0|n).
\end{aligned}$$

By Lemma 6.3, $J_1(\pi')$ is optimized with the choice

$$\hat{\pi}'_i(0) = \hat{\pi}_i(0|n) \qquad \text{(EM−1)}$$

since

$$\sum_{i=1}^{N} \hat{\pi}_i(0|n) = 1\,.$$

Continuing, in a similar way,

$$J_2(D') = \sum_{i=1}^{N} \log\langle D'e_i,\, y_0\rangle\, \hat{\pi}_i(0|n) = \sum_{i=1}^{N}\sum_{r=1}^{M} \log D'_{ri}\, \langle y_0,\, f_r\rangle\, \hat{\pi}_i(0|n)$$

and so for each i, $J_2(D')$ is optimized with the choice

$$D'_{ri} = \langle y_0,\, f_r\rangle \tag{EM−2}$$

which is rather degenerate. However (as remarked in the model of Chapter 9, cf. van der Hoek and Elliott, 2013) if we have observations for L sequences (for example in genomics applications), then the update would be

$$\hat{D}'_{ri} = \frac{\sum_{m=1}^{L} \hat{\pi}_i^m(0|n)\langle y_0^m,\, f_r\rangle}{\sum_{m=1}^{L} \hat{\pi}_i^m(0|n)}$$

where the symbol with the superscript m are values computed for sequence m of the observations for each $1 \leq m \leq L$. The other parameter estimations could similarly be modified for multiple sequence observations.

We now optimize

$$J_3(C') = \mathbf{E}_\theta\left[\log\prod_{l=1}^{n}\langle C'(X_l)y_{l-1}, y_l\rangle \,\middle|\, Y_{0:n} = y_{0:n}\right]$$

$$= \sum_{l=1}^{n} \mathbf{E}_\theta\left[\log\langle C'(X_l)y_{l-1}, y_l\rangle \mid Y_{0:n} = y_{0:n}\right]$$

$$= \sum_{l=1}^{n}\sum_{j=1}^{N}\sum_{r=1}^{M}\sum_{s=1}^{M} \mathbf{E}_\theta\,[\,\langle X_l, e_j\rangle\,\langle y_{l-1}, f_r\rangle\,\langle y_l, f_s\rangle \times$$

$$\times\;\log\langle C'(X_l)y_{l-1}, y_l\rangle \;\mid Y_{0:n} = y_{0:n}]$$

$$= \sum_{l=1}^{n}\sum_{j=1}^{N}\sum_{r=1}^{M}\sum_{s=1}^{M} \log\left(C'^{j}_{sr}\right)\,\langle y_{l-1}, f_r\rangle\,\langle y_l, f_s\rangle\,\mathbf{E}_\theta\left[\langle X_l, e_j\rangle \mid Y_{0:n} = y_{0:n}\right]$$

$$= \sum_{l=1}^{n}\sum_{j=1}^{N}\sum_{r=1}^{M}\sum_{s=1}^{M} \log\left(C'^{j}_{sr}\right)\,\langle y_{l-1}, f_r\rangle\,\langle y_l, f_s\rangle\,\hat{\pi}_j(l|n)$$

and so the update for C is: for each fixed j, r, we optimize over that C'^{j}_{rs} for $1 \leq s \leq M$ to give

$$\hat{C}'^{j}_{sr} = \frac{\sum_{l=1}^{n} \hat{\pi}_j(l|n)\,\langle y_{l-1}, f_r\rangle\,\langle y_l, f_s\rangle}{\sum_{l=1}^{n} \hat{\pi}_j(l|n)\,\langle y_{l-1}, f_r\rangle}\,. \tag{EM−3}$$

If the denominator here is zero, then we do not update this parameter. This comment also applies to the other updates.

We now optimize

$$
\begin{aligned}
J_4(A') &= \mathbf{E}_\theta \left[\log \prod_{l=1}^{n} \langle B'_{l-1}(X_{l-1})X_{l-1}, X_l\rangle \,\middle|\, Y_{0:n} = y_{0:n} \right] \\
&= \sum_{l=1}^{n} \mathbf{E}_\theta \left[\log\langle B'_{l-1}(X_{l-1})X_{l-1}, X_l\rangle \mid Y_{0:n} = y_{0:n} \right] \\
&= \sum_{l=1}^{n}\sum_{j=1}^{N}\sum_{i=1}^{N}\sum_{d=1}^{l} \mathbf{E}_\theta \left[\langle X_l,\, e_j\rangle \, \langle X_{l-1},\, e_i\rangle \, \mathrm{I}(h_{l-1} = d) \right. \\
&\qquad \left. \times \log\langle B'_{l-1}(X_{l-1})X_{l-1}, X_l\rangle \mid Y_{0:n} = y_{0:n} \right] \\
&= \sum_{l=1}^{n}\sum_{j=1}^{N}\sum_{i=1}^{N}\sum_{d=1}^{l} \xi_{l-1}(j,i,d) \log \left\{ \delta_{ij} \frac{F'_i(d+1)}{F'_i(d)} + \frac{p'_i(d)}{F'_i(d)} A'_{ji}(d) \right\} \\
&= \sum_{l=1}^{n}\sum_{j=1, j\neq i}^{N}\sum_{i=1}^{N}\sum_{d=1}^{l} \xi_{l-1}(j,i,d) \log \left\{ \frac{p'_i(d)}{F'_i(d)} A'_{ji}(d) \right\} \\
&\quad + \sum_{l=1}^{n}\sum_{i=1}^{N}\sum_{d=1}^{l} \xi_{l-1}(i,i,d) \log \left\{ \frac{F'_i(d+1)}{F'_i(d)} \right\} \\
&= \sum_{i=1}^{N}\sum_{d=1}^{n}\sum_{j=1, j\neq i}^{N}\sum_{l=d}^{n} \xi_{l-1}(j,i,d) \log A'_{ji}(d) \\
&\quad + \sum_{i=1}^{N}\sum_{d=1}^{n} \lambda_{i,d} \left[\log p'_i(d) - \log F'_i(d) \right] + \mu_{i,d} \left[\log F'_i(d+1) - \log F'_i(d) \right]
\end{aligned}
$$

where we interchange the summations

$$
\sum_{l=1}^{n}\sum_{d=1}^{l} \cdots = \sum_{d=1}^{n}\sum_{l=d}^{n} \cdots
$$

and where

$$
\begin{aligned}
\xi_{l-1}(j,i,d) &= \mathbf{E}_\theta \left[\langle X_l,\, e_j\rangle \, \langle X_{l-1},\, e_i\rangle \, \mathrm{I}(h_{l-1} = d) \mid Y_{0:n} = y_{0:n} \right] \\
\lambda_{i,d} &= \sum_{j=1, j\neq i}^{N} \sum_{l=d}^{n} \xi_{l-1}(j,i,d) \\
\mu_{i,d} &= \sum_{l=d}^{n} \xi_{l-1}(i,i,d).
\end{aligned}
$$

These are quantities which we shall compute later.

It is also clear from this last expression that the updates for $A_{ji}(d)$ and $p_i(d)$ can be made separately. In fact for each i, d we maximize over $j \neq i$ and obtain the update for A

$$\hat{A}'_{ji}(d) = \frac{\sum_{l=d}^{n} \xi_{l-1}(j,i,d)}{\sum_{i=1}^{N} \sum_{i=1}^{N} \sum_{j=1,j\neq i}^{N} \sum_{l=d}^{n} \xi_{l-1}(j,i,d)} \tag{EM-4}$$

To obtain the update for $p_i(d)$, we need the following lemma.

Lemma 10.12 *Let $\alpha, \beta > 0$. Then the function F defined for $x, y > 0$ by*

$$F(x,y) = \alpha \log x - (\alpha+\beta)\log(x+y) + \beta \log y$$

is positive homogeneous and attains a maximum when

$$x = \lambda\,\alpha, \quad y = \lambda\,\beta \quad \textit{where} \quad \lambda > 0.$$

Proof It is clear that

$$F(\lambda\,x, \lambda\,y) = \lambda\,F(x,y)$$

when $\lambda > 0$ and $x, y > 0$. Let

$$z = \frac{x}{x+y} \in (0,1)$$

then

$$F(x,y) = \alpha\,\log z + \beta\,\log(1-z) \equiv f(z).$$

We then have $f'(z) = 0$ implies $z = \alpha/(\alpha+\beta)$ and as $f''(z) < 0$ if $0 < z < 1$, the claim about the maximum of F holds. □

We shall apply this lemma with $\alpha = \lambda_{i,d}$, $\beta = \mu_{i,d}$, $x = p'_i(d)$ and $y = F'_i(d+1)$. With these choices $x + y = F'_i(d)$. For fixed i we wish to optimize

$$J(p'_i) = \sum_{d=1}^{n} \lambda_{i,d}\left[\log p'_i(d) - \log F'_i(d)\right] + \mu_{i,d}\left[\log F'_i(d+1) - \log F'_i(d)\right] \tag{10.3}$$

for $d = 1, 2, \ldots, n$. In updating for the values of $\{p_i(d)\}$ we choose to do this in a non-parametric way.

We now recall some facts. From the definitions given in Lemma 8.3:

$$F_i(d) = \sum_{l=d}^{\infty} p_i(l) = p_i(d) + F_i(d+1) \quad \text{and} \quad F_i(1) = 1\,.$$

To find the update for $p_i(1)$ we maximize

$$\lambda_{i,1}\left[\log p_i'(1) - \log F_i'(1)\right] + \mu_{i,1}\left[\log F_i'(2) - \log F_i'(1)\right]$$

and obtain the update, by Lemma 10.12 this is

$$\hat{p}_i'(1) = \frac{\lambda_{i,1}}{\lambda_{i,1}+\mu_{i,1}} \quad \text{and} \quad \hat{F}_i'(2) = \frac{\mu_{i,1}}{\lambda_{i,1}+\mu_{i,1}}.$$

These sum to 1 as $\hat{F}_i'(1) = 1$.

To find the update for $p_i(2)$ we maximize

$$\lambda_{i,2}\left[\log p_i'(2) - \log F_i'(2)\right] + \mu_{i,2}\left[\log F_i'(3) - \log F_i'(2)\right]$$

and obtain the update, by Lemma 10.12 as

$$\hat{p}_i'(2) = \frac{\lambda_{i,2}}{\lambda_{i,2}+\mu_{i,2}}\left(1-\hat{p}_i'(1)\right) \quad \text{and} \quad \hat{F}_i'(3) = \frac{\mu_{i,2}}{\lambda_{i,2}+\mu_{i,2}}\left(1-\hat{p}_i'(1)\right).$$

These sum to $1-\hat{p}_i'(1)$ as $\hat{F}_i'(2) = 1-\hat{p}_i'(1)$.

We may continue in this way to obtain updates

$$\hat{p}_i'(k) = \frac{\lambda_{i,k}}{\lambda_{i,k}+\mu_{i,k}}\left(1-\hat{p}_i'(1)-\cdots-\hat{p}_i'(k-1)\right)$$

and

$$\hat{F}_i'(k+1) = \frac{\mu_{i,k}}{\lambda_{i,k}+\mu_{i,k}}\left(1-\hat{p}_i'(1)-\cdots-\hat{p}_i'(k-1)\right)$$

for $2 \le k \le n$. This procedure provides non-parametric updates for

$$\hat{p}_i'(1),\ \hat{p}_i'(2),\ \ldots,\ \hat{p}_i'(n) \quad \text{and} \quad \hat{F}_i'(n+1)$$

which is all that one can estimate based on the given observations.

We now express the objective in (10.3) in terms of the updates.

Lemma 10.13 *With parameter θ and $0 \le k \le n-1$, we have*

$$\xi_k(j,i,d) = \frac{1}{L_\theta}\,\beta_{j,1}(k+1)\,A_{ji}(d)\frac{p_i(d)}{F_i(d)}\,\alpha_{i,d}(k)$$

if $j \ne i$ and

$$\xi_k(i,i,d) = \frac{1}{L_\theta}\,\beta_{i,d+1}(k+1)\,\frac{F_i(d+1)}{F_i(d)}\,\alpha_{i,d}(k)$$

where

$$L_\theta = P_\theta\left(Y_{0:n} = y_{0:n}\right)$$

is calculated using Corollary 10.5.

Proof We have for $k+1 \le n$,

$$\xi_k(j,i,d) = \frac{P_\theta\left(X_{k+1}=e_j,\, X_k=e_i,\, h_k=d,\, Y_{0:n}=y_{0:n}\right)}{P_\theta\left(Y_{0:n}=y_{0:n}\right)}.$$

The numerator is

$$\begin{aligned}
&\mathbf{E}_\theta\left[\langle X_{k+1},\, e_j\rangle \langle X_k,\, e_i\rangle \, \mathrm{I}(h_k=d) \prod_{l=0}^{n} \langle Y_l,\, y_l\rangle\right] \\
&= \mathbf{E}_\theta\left[\mathbf{E}_\theta\left[\langle X_{k+1},\, e_j\rangle \langle X_k,\, e_i\rangle \, \mathrm{I}(h_k=d) \prod_{l=0}^{n} \langle Y_l,\, y_l\rangle \,\middle|\, \mathcal{F}_n\right]\right] \\
&= \mathbf{E}_\theta\left[\langle X_{k+1},\, e_j\rangle \langle X_k,\, e_i\rangle \, \mathrm{I}(h_k=d)\, \mathbf{E}_\theta\left[\prod_{l=0}^{n} \langle Y_l,\, y_l\rangle \,\middle|\, \mathcal{F}_n\right]\right] \\
&= \mathbf{E}_\theta\left[\langle X_{k+1},\, e_j\rangle \langle X_k,\, e_i\rangle \, \mathrm{I}(h_k=d) \langle D\, X_0,\, y_0\rangle \prod_{l=1}^{n} \langle C(X_l)\, y_{l-1},\, y_l\rangle\right] \\
&= \mathbf{E}_\theta\Bigg[\langle X_{k+1},\, e_j\rangle \langle X_k,\, e_i\rangle \, \mathrm{I}(h_k=d) \langle D\, X_0,\, y_0\rangle \prod_{l=1}^{k} \langle C(X_l)\, y_{l-1},\, y_l\rangle \\
&\qquad \times \mathbf{E}_\theta\left[\prod_{l=k+1}^{n} \langle C(X_l)\, y_{l-1},\, y_l\rangle \,\middle|\, \mathcal{F}_{k+1}\right]\Bigg] \\
&= \mathbf{E}_\theta\Bigg[\langle X_{k+1},\, e_j\rangle \langle X_k,\, e_i\rangle \, \mathrm{I}(h_k=d) \langle D\, X_0,\, y_0\rangle \prod_{l=1}^{k} \langle C(X_l)\, y_{l-1},\, y_l\rangle \\
&\qquad \times \mathbf{E}_\theta\left[\prod_{l=k+1}^{n} \langle C(X_l)\, y_{l-1},\, y_l\rangle \,\middle|\, X_{k+1},\, h_{k+1}\right]\Bigg].
\end{aligned}$$

We now split the argument into two cases.

Case 1: For $j=i$ and $h_{k+1}=d+1$.
The last expression becomes

$$\begin{aligned}
&\beta_{i,d+1}(k+1)\mathbf{E}_\theta\Bigg[\langle X_{k+1}, e_i\rangle \langle X_k, e_i\rangle \mathrm{I}(h_k=d) \times \\
&\qquad\qquad \times \; \langle D\, X_0, y_0\rangle \prod_{l=1}^{k} \langle C(X_l) y_{l-1}, y_l\rangle\Bigg] \\
&= \beta_{i,d+1}(k+1)\, \mathbf{E}_\theta\left[\mathbf{E}_\theta\left[\text{ditto} \mid \mathcal{F}_k\right]\right]
\end{aligned}$$

$$= \beta_{i,d+1}(k+1)\mathbf{E}_\theta\Big[\langle X_k, e_i\rangle \, \mathrm{I}(h_k = d)\, \langle D\, X_0,\, y_0\rangle \times$$

$$\times \prod_{l=1}^{k} \langle C(X_l)\, y_{l-1},\, y_l\rangle \mathbf{E}_\theta\Big[\langle X_{k+1},\, e_i\rangle \Big| \mathcal{F}_k\Big]\Big]$$

$$= \beta_{i,d+1}(k+1)\, \frac{F_i(d+1)}{F_i(d)}\, \mathbf{E}_\theta\Big[\langle X_k,\, e_i\rangle \, \mathrm{I}(h_k = d)\, \langle D\, X_0,\, y_0\rangle \times$$

$$\times \prod_{l=1}^{k} \langle C(X_l)\, y_{l-1},\, y_l\rangle\Big]$$

$$= \beta_{i,d+1}(k+1)\, \frac{F_i(d+1)}{F_i(d)}\, \mathbf{E}_\theta\left[\langle X_k,\, e_i\rangle \, \mathrm{I}(h_k = d) \prod_{l=0}^{k} \langle Y_l\,,\, y_l\rangle\right]$$

$$= \beta_{i,d+1}(k+1)\, \frac{F_i(d+1)}{F_i(d)}\, \alpha_{i,d}(k).$$

Case 2: For $j \neq i$ and $h_{k+1} = 1$.

The last expression then becomes in a similar way

$$\beta_{i,1}(k+1)\mathbf{E}_\theta\Big[\langle X_k, e_i\rangle \mathrm{I}(h_k = d)\langle DX_0, y_0\rangle \times$$

$$\times \prod_{l=1}^{k} \langle C(X_l) y_{l-1}, y_l\rangle \mathbf{E}_\theta\Big[\langle X_{k+1}, e_j\rangle \Big| \mathcal{F}_k\Big]\Big]$$

$$= \beta_{i,1}(k+1)\, A_{ji}(d)\, \frac{p_i(d)}{F_i(d)}\, \mathbf{E}_\theta\left[\langle X_k,\, e_i\rangle \, \mathrm{I}(h_k = d) \prod_{l=0}^{k} \langle Y_l, y_l\rangle\right]$$

$$= \beta_{i,1}(k+1)\, A_{ji}(d)\, \frac{p_i(d)}{F_i(d)}\, \alpha_{i,d}(k).$$

and the lemma is proved. □

Appendix A
Higher-Order Chains

In this appendix we provide a further discussion of higher-order chains.

Recall the definition of the matrix A when $M = 2$ and $N = 2$:

$$A = \begin{bmatrix} a_{1,11} & a_{1,12} & a_{1,21} & a_{1,22} \\ a_{2,11} & a_{2,12} & a_{2,21} & a_{2,22} \end{bmatrix}.$$

Here

$$a_{k,ji} = P[\,X_{n+1} = e_k \,|\, X_n = e_j, X_{n-1} = e_i].$$

We wish to construct the matrix Π

$$\Pi = \begin{bmatrix} a_{1,11} & a_{1,12} & 0 & 0 \\ 0 & 0 & a_{1,21} & a_{1,22} \\ a_{2,11} & a_{2,12} & 0 & 0 \\ 0 & 0 & a_{2,21} & a_{2,22} \end{bmatrix}.$$

Taking Kronecker products in lexicographic order, the standard unit vectors in $\mathbb{R}^4$ are:

$$\begin{aligned} f_1 &= e_1 \otimes e_1 \\ f_2 &= e_1 \otimes e_2 \\ f_3 &= e_2 \otimes e_1 \\ f_4 &= e_2 \otimes e_2\,. \end{aligned}$$

Also recall the definition of Z_n

$$Z_n = X_n \otimes X_{n-1}\,.$$

Therefore,

$$\pi_{j,i} = P[\,Z_{n+1} = f_j \,|\, Z_n = f_i].$$

Note that some of the elements of $\Pi = (\pi_{j,i})$ are zero. For example, $\pi_{1,3} = 0$. This is because $f_3 \to f_1$ is not possible, that is,

$$f_3 = e_2 \otimes e_1 \nrightarrow f_1 = e_1 \otimes e_1$$

We now show how to write Π in terms of A. Firstly we define A_1 and A_2 in the following way

$$A_1 = A \begin{bmatrix} 1 & 0 & 0 & 0 \\ 0 & 1 & 0 & 0 \\ 0 & 0 & 0 & 0 \\ 0 & 0 & 0 & 0 \end{bmatrix} = \begin{bmatrix} a_{1,11} & a_{1,12} & 0 & 0 \\ a_{2,11} & a_{2,12} & 0 & 0 \end{bmatrix}$$

$$A_2 = A \begin{bmatrix} 0 & 0 & 0 & 0 \\ 0 & 0 & 0 & 0 \\ 0 & 0 & 1 & 0 \\ 0 & 0 & 0 & 1 \end{bmatrix} = \begin{bmatrix} 0 & 0 & a_{1,21} & a_{1,22} \\ 0 & 0 & a_{2,21} & a_{2,22} \end{bmatrix}.$$

Pre-multiply A_1 and A_2 by the following matrices

$$\begin{bmatrix} 1 & 0 \\ 0 & 0 \\ 0 & 1 \\ 0 & 0 \end{bmatrix} A_1 = \begin{bmatrix} a_{1,11} & a_{1,12} & 0 & 0 \\ 0 & 0 & 0 & 0 \\ a_{2,11} & a_{2,12} & 0 & 0 \\ 0 & 0 & 0 & 0 \end{bmatrix}$$

$$\begin{bmatrix} 0 & 0 \\ 1 & 0 \\ 0 & 0 \\ 0 & 1 \end{bmatrix} A_2 = \begin{bmatrix} 0 & 0 & 0 & 0 \\ 0 & 0 & a_{1,21} & a_{1,22} \\ 0 & 0 & 0 & 0 \\ 0 & 0 & a_{2,21} & a_{2,22} \end{bmatrix}.$$

Then finally, we obtain

$$\Pi = \begin{bmatrix} 1 & 0 \\ 0 & 0 \\ 0 & 1 \\ 0 & 0 \end{bmatrix} A_1 + \begin{bmatrix} 0 & 0 \\ 1 & 0 \\ 0 & 0 \\ 0 & 1 \end{bmatrix} A_2 .$$

Then

$$A_1 = A\ (\text{diag}(e_1) \otimes I_2) \quad \text{where} \quad \text{diag}(e_1) = \begin{bmatrix} 1 & 0 \\ 0 & 0 \end{bmatrix}$$

$$A_2 = A\ (\text{diag}(e_2) \otimes I_2) \quad \text{where} \quad \text{diag}(e_2) = \begin{bmatrix} 0 & 0 \\ 0 & 1 \end{bmatrix}.$$

Notice that $A_1 + A_2 = AI_4 = A$. We can use this decomposition of A and pre-multipliers to reconstruct Π in a different way. Thus

$$\begin{aligned}\Pi &= (\mathrm{I}_2 \otimes e_1)\, A_1 + (\mathrm{I}_2 \otimes e_2)\, A_2 \\ &= \sum_{i=1}^{2} (\mathrm{I}_2 \otimes e_i)\, A\, (\mathrm{diag}(e_i) \otimes \mathrm{I}_2).\end{aligned}$$

Write $\mathrm{I}_2 \otimes e_i = \mathbb{K}_i$ and $\mathrm{diag}(e_i) \otimes I_2 = \mathbb{I}_i$, so that

$$\Pi = \sum_{i=1}^{2} \mathbb{K}_i\, A\, \mathbb{I}_i.$$

This form will eventually generalize to general M and N.

We can also write

$$J_1 = \begin{bmatrix} 1 & 1 & 0 & 0 \\ 0 & 0 & 1 & 1 \end{bmatrix} = I_2 \otimes [1, 1]$$

and

$$J_2 = \begin{bmatrix} 1 & 0 & 1 & 0 \\ 0 & 1 & 0 & 1 \end{bmatrix} = [1, 1] \otimes I_2\,.$$

Again this concept will be generalized.

Note that

$$J_1\, \Pi = \sum_{i=1}^{2} J_1 \mathbb{K}_i\, A\, \mathbb{I}_i\,.$$

As

$$J_1 \mathbb{K}_i = (\mathrm{I}_2 \otimes [1, 1])(\mathrm{I}_2 \otimes e_i) = (\mathrm{I}_2 \cdot I_2 \otimes [1, 1] e_i) = \mathrm{I}_2 \otimes 1 = \mathrm{I}_2$$

and

$$\mathrm{I}_4 = \mathbb{I}_1 + \mathbb{I}_2$$

then

$$J_1\, \Pi = \sum_{i=1}^{2} \mathrm{I}_2\, A\, \mathbb{I}_i = \sum_{i=1}^{2} A\, \mathbb{I}_i = A \sum_{i=1}^{2} \mathbb{I}_i = A\, \mathrm{I}_4 = A.$$

In these calculations we used the following property of the Kronecker product:

$$(A \otimes B)(C \otimes D) = AC \otimes BD$$

which holds whenever $A\,C$ and $B\,D$ are defined. This formula has a generalization

$$(A_1 \otimes A_2 \otimes \cdots \otimes A_k)\, (B_1 \otimes B_2 \otimes \cdots \otimes B_k) = A_1\, B_1 \otimes \cdots \otimes A_k\, B_k \quad \text{(A.1)}$$

provided all the products are defined. We shall use this generalization a number of times below (Steen and Hardy, 2011).

It is also useful to write $A = (A_{j,i})$ as well in the former notation with

$$A_{j,i} = P[\,X_{n+1} = e_j \mid Z_n = f_i\,]$$

and where for example $a_{1,21} = A_{1,3}$ since $e_2 \otimes e_1 = f_3$, and so on.

For programming purposes, it may be useful to have a formula for k when

$$f_k = e_i \otimes e_j.$$

In this case where $M = N = 2$ the formula is

$$k = 2 \cdot (i-1) + j\,.$$

This can be easily checked here

$$\begin{array}{lllllll} f_1 & = & e_1 \otimes e_1 & \text{and} & 2(1-1)+1 & = & 1 \\ f_2 & = & e_1 \otimes e_2 & \text{and} & 2(1-1)+2 & = & 2 \\ f_3 & = & e_2 \otimes e_1 & \text{and} & 2(2-1)+1 & = & 3 \\ f_4 & = & e_2 \otimes e_2 & \text{and} & 2(2-1)+2 & = & 4. \end{array}$$

Conversely, given f_k then

$$\begin{array}{lll} e_i & = & J_1 f_k \\ e_j & = & J_2 f_k\,. \end{array}$$

As all transitions are not possible, for a given f_j we would like to determine which f_k may follow it. For $f_3 = e_2 \otimes e_1$, the answer is easily determined from this decomposition of f_3. The possibilities are $f_2 = e_1 \otimes e_2$ or $f_4 = e_2 \otimes e_2$. Without making the decomposition, the answer is

$$\frac{1}{2}[\,(k-1) - (k-1)\,\text{modulo}\,2] + 1 + 2\,i$$

for $i = 0, 1$. This formula will make more sense if we generalize it to general $M \geq 2$ and $N \geq 2$. For $k = 3$ we have

$$\frac{1}{2}[\,(3-1) - (3-1)\,\text{modulo}\,2] + 1 + 2\,i = 1 + 1 + 2\,i$$

for $i = 0, 1$ which gives labels 2 and 4, which is correct. Note that for any f_k there are only two followers.

Appendix B
An Example of a Second-Order Chain

An example of a second-order chain, $M = 2$, using a 2-letter alphabet, $N = 2$

Consider the following sequence using a 2-letter alphabet.

```
(end) 3' B B A B A B B A A A B A B 5' (start)
e_i      2 2 1 2 1 2 2 1 1 1 2 1 2
f_j      4 3 2 3 2 4 3 1 1 2 3 2 -
```

We use this as data to calibrate a second-order Markov chain. Therefore, with $\Pi = (\pi_{ji})$ as in Appendix A, we estimate (reading **right to left**)

$$\hat{\pi}_{4,3} = \frac{\text{the number of times 4 follows 3}}{\text{the number of times 3 occurs}} = \frac{2}{4} = 0.5$$

$$\hat{\pi}_{2,3} = \frac{\text{the number of times 2 follows 3}}{\text{the number of times 3 occurs}} = \frac{2}{4} = 0.5$$

and $\pi_{1,3} = \pi_{3,3} = 0$.

This is a sample of length $L = 12$. When calculating the number of times 4 occurs in the sequence, this form fits $L-1$ terms in the sequence. So

$$\hat{\pi}_{2,4} = \frac{\text{the number of times 2 follows 4}}{\text{the number of times 4 occurs}} = \frac{1}{1} = 1.$$

The rest of Π is estimated in the same way. We obtain

$$\widehat{\Pi} = \begin{bmatrix} 0.5 & 0.25 & 0 & 0 \\ 0 & 0 & 0.5 & 1 \\ 0.5 & 0.75 & 0 & 0 \\ 0 & 0 & 0.5 & 0 \end{bmatrix}.$$

We can then use this matrix to obtain the estimate for the matrix A,

$$\widehat{A} = J_1 \widehat{\Pi} = \begin{bmatrix} 0.5 & 0.25 & 0.5 & 1 \\ 0.5 & 0.75 & 0.5 & 0 \end{bmatrix}.$$

The general case

We now consider the general case of $N \geq 2$ and $M \geq 2$. Here $\mathbb{R}^N$ is the state space of the chain X and M is the order of the Markov chain.

For $n \geq M$, we **define** $Z_n = X_n \otimes X_{n-1} \otimes \cdots \otimes X_{n-M+1}$. (An alternative definition is to write this Knonecker product in reverse order as in Aggoun and Elliott (2004), on pages 58–59 of which there is a brief discussion of a second-order Markov chain, with this alternative definition.)

The state space of Z_n is now $\mathbb{R}^{N^M}$ or more specifically the collection

$$\{f_k \; : \; k = 1, 2, \ldots, N^M\}$$

where as usual each f_k has the form

$$(0, 0, 0, \ldots, 1, 0, \ldots .0)'$$

with 1 in the kth coordinate.

Lemma B.1 *Each f_k can be represented as*

$$f_k = e_{i_1} \otimes e_{i_2} \otimes \cdots \otimes e_{i_m}$$

and k is given by

$$k = \sum_{l=1}^{M} (i_l - 1) N^{M-l} + 1. \tag{B.1}$$

We leave its proof as an exercise.

Example When $M = 2$ and $N = 2$ then $e_2 \otimes e_1 = f_3$ as

$$(2-1)2^{2-1} + (1-1)2^{2-2} + 1 = 3.$$

Lemma B.2 *If $Z_n = f_j$, then there are N possible choices of $Z_{n+1} = f_k$ where*

$$k = (l-1)N^{M-1} + \frac{1}{N}\{(j-1) - (j-1)\, modulo\, N\} + 1$$

for $l = 1, 2, \ldots, N$.

Proof If

$$Z_n = f_j = e_{i_1} \otimes e_{i_2} \cdots \otimes e_{i_M} \tag{B.2}$$

then

$$Z_{n+1} = f_k = e_l \otimes e_{i_1} \otimes \cdots \otimes e_{i_{M-1}} \quad \text{for } l = 1, \ldots, N.$$

Note only the first term here is a new factor. We now employ Lemma B.1 so if Z_n is the f_j of (B.2)

$$j = \sum_{l=1}^{M} (i_l - 1) N^{M-l} + 1.$$

This implies that

$$k = (l-1) N^{M-1} + \frac{1}{N}(j - i_M) + 1.$$

If we use

$$i_M = (j-1) \bmod N + 1$$

the result follows. □

Example For the choice $M = 2, N = 2, j = 3, i_M = (3-1) \bmod 2 + 1 = 1$ and so possible values of k are

$$k = (l-1)\, 2^1 + \frac{1}{2}(3-1) + 1 \quad \text{for } l = 0, 1.$$

Then we have $k = 2$ and $k = 4$ as before.

Definition B.3 For $i = 1, 2, \ldots, M$, define

$$J_i = 1_N^\top \otimes 1_N^\top \otimes \cdots \otimes \mathrm{I}_N \otimes 1_N^\top \otimes \cdots \otimes 1_N^\top$$

where I_N is the $N \times N$ identity matrix, $1_N^\top = (1, 1, \ldots, 1)$ (N ones). The matrix J_i is the Kronecker product of M terms and the I_N in this expression is in the ith term.

The matrix J_i is $N \times N^M$ in size for each i.

Lemma B.4 *If for* $n \geq M$

$$Z_n = X_n \otimes X_{n-1} \otimes \cdots \otimes X_{n-M+1}$$

then

$$X_{n-j+1} = J_i\, Z_n$$

for each $j = 1, 2, \ldots, M$.

Proof It will suffice to give a proof for $j = 1$, $j = 2$ and for

$$Z_n = X_n \otimes X_{n-1} \otimes \cdots \otimes X_{n-M+1} = e_{i_1} \otimes e_{i_2} \cdots \otimes e_{i_M} .$$

Then

$$J_1 \, Z_n = \left(\mathrm{I}_N \otimes 1_N^\top \otimes \cdots \otimes 1_N^\top\right) \left(e_{i_1} \otimes e_{i_2} \otimes \cdots \otimes e_{i_M}\right)$$

and by (A.1),

$$\begin{aligned} J_1 \, Z_n &= \mathrm{I}_N e_{i_1} \otimes 1_N^\top e_{i_2} \otimes 1_N^\top e_{i_3} \otimes \cdots \otimes 1_N^\top e_{i_M} \\ &= e_{i_1} \otimes 1 \otimes 1 \otimes \cdots \otimes 1 = e_{i_1} = X_n . \end{aligned}$$

Similarly

$$\begin{aligned} J_2 \, Z_n &= 1_N^\top e_{i_1} \otimes \mathrm{I}_N e_{i_2} \otimes 1_N^\top e_{i_3} \otimes \cdots \otimes 1_N^\top e_{i_M} \\ &= 1 \otimes e_{i_2} \otimes 1 \otimes \cdots \otimes 1 = e_{i_2} = X_{n-1} . \end{aligned}$$

□

Definition B.5 For $i = 1, 2, \ldots, N^{M-1}$, let g_i be basis elements of $\mathbb{R}^{N^{M-1}}$ of the form

$$(0, 0, 0, \ldots, 1, 0, \ldots .0)^\top$$

with 1 in the kth coordinate.

For each $i = 1, 2, \ldots, N^{M-1}$, let

$$\mathbb{K}_i = \mathrm{I}_N \otimes g_i$$

which is an $N^M \times N$ matrix.

For each $i = 1, 2, \ldots, N^{M-1}$, let

$$\mathbb{I}_i = \mathrm{diag}\,(g_i) \otimes \mathrm{I}_N$$

which is an $N^M \times N^M$ matrix.

In this definition by $\mathrm{diag}\,(\mathbf{c})$ we mean for $\mathbf{c} \in \mathbb{R}^p$, the $p \times p$ matrix C with zero off-diagonal elements and $C_{i,i} = c_i$ for $i = 1, 2, \ldots, p$.

Lemma B.6 *We have the following representation of* Π

$$\Pi = \sum_{i=1}^{N^{M-1}} \mathbb{K}_i \, A \, \mathbb{I}_i .$$

Remark We believe that this is a new representation of Π and allows for the easy description of a Markov chain of order $M \geq 2$ as a first-order Markov chain with a modified state space. It is motivated by the initial results in this direction by Paul Malcolm cited earlier.

We note that

$$\begin{aligned} J_1\,\mathbb{K}_i &= \left(\mathrm{I}_N \otimes 1_N^\top \otimes \ldots \otimes 1_N^\top\right)\left(\mathrm{I}_N \otimes g_i\right) \\ &= \mathrm{I}_N \otimes \left(\left(1_N^\top \otimes \ldots \otimes 1_N^\top\right) g_i\right) = \mathrm{I}_N \otimes 1 = \mathrm{I}_N \end{aligned}$$

since the $M-1$ times Kronecker product

$$1_N^\top \otimes \cdots \otimes 1_N^\top = (1, 1, \ldots, 1),$$

with N^{M-1} terms which are all 1.

Also note that

$$\mathrm{I}_{N^M} = \sum_{i=1}^{N^{M-1}} \mathbb{I}_i$$

and so it follows that

$$J_1\,\Pi = A.$$

Appendix C
A Conditional Bayes Theorem

We repeatedly used the Bayes' conditional expectation formula (Elliott et al., 1995, Theorem 3.2, p. 22). We summarize a proof:

- Let $\Lambda \geq 0$ be a random variable on $(\Omega, \mathcal{F}, \mathcal{P})$ with $\mathbf{E}[\Lambda] = 1$.
- Let $Q : \mathcal{F} \to [0, 1]$ be defined by $Q(A) = \mathbf{E}[\Lambda I(A)]$ for all $A \in \mathcal{F}$. Clearly Q is a probability on $(\Lambda, \mathcal{F})$.
- Let $\mathbf{E}, \mathbf{E}^Q$ be expectations with respect to P and Q, respectively.
- Let X be a random variable on $(\Omega, \mathcal{F}, \mathcal{P})$ with $\mathbf{E}^Q[|X|] = \mathbf{E}[\Lambda|X|] < \infty$.
- Let $\mathcal{G}$ be a σ-algebra contained in $\mathcal{F}$.
- Let $B = \{\omega \in \Omega \mid \mathbf{E}[\Lambda|\mathcal{G}](\omega) = 0\}$. Then $B \in \mathcal{G}$. It could be the case that $P(B) > 0$. However

$$Q(B) = \mathbf{E}[\Lambda I(B)] = \mathbf{E}\left[\mathbf{E}\left[\Lambda I(B) \mid \mathcal{G}\right]\right] = \mathbf{E}\left[I(B)\,\mathbf{E}\left[\Lambda \mid \mathcal{G}\right]\right] = 0.$$

Lemma C.1 (Conditional Bayes Theorem) *We have*

$$\mathbf{E}^Q[X|\mathcal{G}] = \begin{cases} \frac{\mathbf{E}[\Lambda X|\mathcal{G}]}{\mathbf{E}[\Lambda|\mathcal{G}]} & \text{if } \mathbf{E}[\Lambda|\mathcal{G}] > 0 \\ \\ 0 & \text{otherwise} \end{cases} \tag{C.1}$$

Proof The right-hand side of (C.1) is $\mathcal{G}$-measurable, and can be written as

$$\frac{\mathbf{E}[\Lambda X|\mathcal{G}]}{\mathbf{E}[\Lambda|\mathcal{G}]}\,\mathrm{I}(B^c).$$

Let $A \in \mathcal{G}$ be arbitrary.

Then

$$\mathbf{E}^Q\left[I(A)\frac{\mathbf{E}[\Lambda X|\mathcal{G}]}{\mathbf{E}[\Lambda|\mathcal{G}]}\,\mathrm{I}(B^c)\right] = \mathbf{E}\left[\Lambda I(A)\frac{\mathbf{E}[\Lambda X|\mathcal{G}]}{\mathbf{E}[\Lambda|\mathcal{G}]}\,\mathrm{I}(B^c)\right]$$

$$
\begin{aligned}
&= \mathbf{E}\left[\mathbf{E}\left[\Lambda I(A)\frac{\mathbf{E}[\Lambda X|\mathcal{G}]}{\mathbf{E}[\Lambda|\mathcal{G}]}\,\mathrm{I}(B^c)\middle|\,\mathcal{G}\right]\right] \\
&= \mathbf{E}\left[I(A)\frac{\mathbf{E}[\Lambda X|\mathcal{G}]}{\mathbf{E}[\Lambda|\mathcal{G}]}\,\mathrm{I}(B^c)\mathbf{E}[\Lambda|\mathcal{G}]\right] \\
&= \mathbf{E}[I(A)\mathbf{E}[\Lambda X|\mathcal{G}]] \\
&= \mathbf{E}[I(A)\Lambda X] \\
&= \mathbf{E}[\Lambda[I(A)X]] \\
&= \mathbf{E}^Q[I(A)X]
\end{aligned}
$$

and the conditions for a conditional expectation. □

Appendix D
On Conditional Expectations

Let X and Y be random variables with $\mathbf{E}[|X|] < \infty$ on some probability space $(\Omega, \mathcal{F}, P)$. In fact, let us suppose that $X \geq 0$.

By the Doob–Dynkin Lemma (see for example, Oksendal, 2010, Lemma 2.1.2)

$$\mathbf{E}[X|Y] = \mathbf{E}[X|\sigma(Y)] = g(Y)$$

for some Borel function $g : \mathbb{R}^n \to \mathbb{R}$ if Y takes values in $\mathbb{R}^n$.

If Y takes countably many distinct values $\{a_1, a_2, \ldots\}$, then g can be described in more detail. The atoms of $\sigma(Y)$, the σ-algebra generated by Y, are the disjoint sets

$$A_k = \{\omega \in \Omega \,|\, Y(\omega) = a_k\}$$

This means that any element of $\sigma(Y)$ is either $\emptyset$ or a countable union of atoms. The function g is characterized as follows: g is constant on each atom, and

$$\mathbf{E}[g(Y)\,\mathrm{I}(A)] = \mathbf{E}[X\,\mathrm{I}(A)] \tag{D.1}$$

for all $A \in \sigma(Y)$. In fact (D.1) is equivalent to the result holding for atoms A. This leads to

$$g_k\,P(A_k) = \mathbf{E}[X\,\mathrm{I}(A_k)] \tag{D.2}$$

where $g_k = g(a_k)$. If $P(A_k) > 0$, then

$$g_k = \frac{\mathbf{E}[X\,\mathrm{I}(A_k)]}{P(A_k)}\,. \tag{D.3}$$

But if $P(A_k) = 0$, then (D.2) becomes

$$g_k\,P(A_k) = 0$$

as $\mathbf{E}[X\,\mathrm{I}(A_k)] = 0$. This means that g_k can be given any finite value. If we set $g_k = 0$, then

$$\mathbf{E}[X|Y = a_k] = g(a_k) = g_k = 0.$$

So we can say then when $P(A_k) = 0$, then $\mathbf{E}[X|Y = a_k] = 0$.

We could also note that

$$\mathbf{E}[X|Y] = \sum_k g_k\,\mathrm{I}(A_k)$$

and we can set $g_k = 0$ on sets A_k when $P(A_k) = 0$ and obtain a random variable equal to the conditional expected value $\mathbf{E}[X|Y]$ with probability 1. As many authors use this convention, we shall also adopt it.

Now let W be another random variable which takes countably many distinct values $\{b_1, b_2, b_3, \ldots\}$. If $P(Y = a_k) = 0$, then $P(Y = a_k,\, W = b_l) = 0$ for all $l = 1, 2, 3, \ldots$. Then

$$\mathbf{E}[X|Y = a_k, W] = \mathbf{E}[X|Y = a_k] = 0$$

with probability 1. This is because

$$\begin{aligned}&\mathbf{E}[X|Y = a_k, W]\\&\quad = \sum_{l=1}^{\infty} \mathbf{E}[X|Y = a_k, W = b_l]\,\mathrm{I}(Y = a_k,\, W = b_l) = \mathbf{E}[X|Y = a_k]\end{aligned}$$

with probability 1, since

$$P\left(\bigcup_{l=1}^{\infty}\{Y = a_k, W = b_l\}\right) = P\,(Y = a_k) = 0.$$

We could apply these ideas with $X = \mathrm{I}(Z_{n+1} = e_j)$, $Y = (Z_n, T_{n+1} - T_n)$ and $a_k = (e_i, m)$ and $W = (Z_{0:n-1}, T_{0:n})$. If $P(Y = a_k) = 0$, then

$$P(Z_{n+1} = e_j \,|\, Y = a_k,\, W) = P(Z_{n+1} = e_j \,|\, Y = a_k)$$

with probability 1. In the application above, if

$$P(T_{n+1} - T_n = k \,|\, Z_n = e_i) = 0$$

implies

$$P(T_{n+1} - T_n = k,\, Z_n = e_i) = 0.$$

We could also discuss what is meant by $P(A|B)$ when $P(B) = 0$. It should be the value of

$$\mathbf{E}[\mathrm{I}(A)|\sigma(B)]$$

on the set B. Here $\sigma(B) = \{\emptyset, B, B^c, \Omega\}$. With our convention, we would set $P(A|B) = 0$.

Appendix E
Some Molecular Biology

In 1865, the Augustinian friar Gregor Mendel discovered the basic laws of genetic inheritance which are now known as Mendelian Inheritance. Deoxyribonucleic acid (DNA) was isolated in 1869 by Friedrich Miescher. In 1944 Avery, Macleod and McCarty established that it is genetic material. DNA is a nucleic acid containing the genetic instructions used in the development and functioning of all known living organisms. The DNA segments carrying this genetic information are called genes. In 1953, Watson and Crick discovered the molecular structure of DNA.

Genes can be studied on a molecular, cellular, organismal, population, or evolutionary level. DNA is a double helix of two intertwined and complementary nucleotide chains. The entire set of DNA is the **genome** of the organism and the DNA molecules in the genome are assembled into **chromosomes**. Genes are the functional regions of DNA.

Each gene contains information about the structure and behavior of some protein produced in the cell. However, proteins are the mechanism in a cell and they prescribe its behaviour. Proteins play a number of roles such as catalyzing reactions, transporting oxygen and regulating the production of other proteins. Proteins are encoded by genes using two steps: transcription and translation. **Transcription** is the process of copying the information encoded in the DNA into a molecule called messenger ribonucleic acid (mRNA). Many copies of the same RNA can be produced from a single copy of DNA. This copy allows the cell to make large amounts of proteins. This process is referred to as **translation**, and converts the mRNA into chains of linked amino acids called **polypeptides**. Polypeptides can combine with other polypeptides or act on their own, to make the actual proteins. The information flow from DNA to RNA to protein is known as the **central dogma** of molecular biology. However, there are a number of modifications that need to be made. These include reverse transcription, RNA editing and RNA replication.

(Shmulevich and Dougher, 2007, p. 2) provides details on the DNA structure:

- The DNA molecule is a polymer that is strung together from **nucleotides**, each of which consists of three chemical components: a sugar (deoxyribose), a phosphate group, and a nitrogenous base. There are four possible bases: adenine, guanine, cytosine and thymine, often abbreviated to `A`, `G`, `C`, `T`, respectively. Adenine and guanine are **purines** and have bicyclic structures (two fused rings), whereas cytosine and thymine are **pyrimidines**, and have monocyclic structures. The sugar has five carbon atoms that are typically numbered $1'$ to $5'$. The phosphate group is attached to the $5'$-carbon atom, whereas the base is attached to the $1'$-carbon. The $3'$-carbon also has, a hydroxyl group (OH) attached to it.
- The genetic code is triplet based: there are three possible ways a particular message can be read.
- The relationship between the nucleotide sequences of genes and the amino-acid sequences of proteins is determined by the rules of translation, known collectively as the genetic code. The genetic code consists of three-letter 'words' called codons formed from a sequence of three nucleotides. Since there are four bases in 3-letter combinations, we have $4^3 = 64$ possible codons.
- The open reading frame (ORF) includes the initiation codon and the sequence of nucleotides between the start (initiation) codon and an in-frame stop (termination) codon that codes for amino acids to be incorporated.
- The 64 possible codons encode the 20 amino acids (which are also called residues). Shmulevich and Dougher (2007) simply state that several codons may encode the same amino acid and they provide the table (given with `U` replacing `T` as in the RNA molecule). (The 3- and 1-letter codes for amino acids are available in Waterman (2000) and elsewhere):

  We note that some triple letter codons appear here. (Waterman, 2000, p. 10), gives another explanation for these matters but this need not concern us. All that matters is that we know to what a codon encodes. The information stored in DNA is ultimately transferred to proteins. Proteins are linear chains of amino acids and consist of 20 amino acids.

- RNA is like DNA but with single-stranded and different letters {A, U, G, C}. The bases are similar with DNA except thymine where (T) is replaced with uracil (U).

- To conclude, the DNA is like a word which uses the alphabet {A, T, C, G} of nucleotides and RNA is like a word that uses the alphabet {A, U, C, G} of nucleotides. The protein is then like a word that uses an alphabet of 20 amino acids.
- Cells in our world come in two basic types, prokaryotes and eukaryotes. Karyose comes from a Greek word which means 'nut' or 'kernel'. In biology, we use this word root to refer to the nucleus of a cell. *Pro* means 'before' and *eu* means 'true' or 'good'. So prokaryotes means 'before a nucleus' and eukaryotes means 'possessing a true nucleus'. In other words, prokaryotes cells have no nucleus, while eukaryotes cells do have a true nucleus. However, this is far from the only difference between these two cell types – or how humans are different from bacteria and protozoa.

Below we suggest a model for the sequence alignment using Markov chains and hidden Markov models.

The results in later sections of the book provide algorithms for the annotation of the genome, searching for coding regions as well as the genes. These models could also be used to distinguish true genes from false genes and to identify coding and non-coding regions in a gene.

Simple independent, identically distribution models

We now describe a simple model with state space $\mathcal{Q} = \{\mathtt{A}, \mathtt{G}, \mathtt{T}, \mathtt{C}\}$. Consider a simple independent identically distributed (iid) chain model.

A chain is just a sequence

$$\{X_k : k = 1, 2, \ldots, L\}$$

where $X_k \in \mathcal{Q} = \{\mathtt{A}, \mathtt{G}, \mathtt{T}, \mathtt{C}\}$ – the genetic alphabet.

In an iid model the terms of the sequence are independent and have the same distribution specified by

$$P(X_k = \mathtt{A}) = P_\mathtt{A}\,,\ P(X_k = \mathtt{G}) = P_\mathtt{G}\,,\ P(X_k = \mathtt{T}) = P_\mathtt{T}\,,\ P(X_k = \mathtt{C}) = P_\mathtt{C}$$

for each $k = 1, 2, \ldots, L$ and where $P_\mathtt{A}$, $P_\mathtt{G}$, $P_\mathtt{T}$ and $P_\mathtt{C}$ are all positive and

$$P_\mathtt{A} + P_\mathtt{G} + P_\mathtt{T} + P_\mathtt{C} = 1.$$

Of course the values $P_{\mathtt{A}}$, $P_{\mathtt{G}}$, $P_{\mathtt{T}}$ and $P_{\mathtt{C}}$ are unknown and need to be estimated.

In this book we provide algorithms to estimate or calibrate models based on observations. We would expect that the answers for the estimated $\widehat{P}_{\mathtt{A}}^L$ based on sequences of length L should be

$$\widehat{P}_{\mathtt{A}}^L = \frac{1}{L}\sum_{k=1}^{L} I(X_k = \mathtt{A})$$

and we should expect that $\widehat{P}_{\mathtt{A}}^L \to P_{\mathtt{A}}$ in some sense as $L \to \infty$.

Here we used the notation

$$I(X = A) = \begin{cases} 1 & \text{if } X = \mathtt{A} \\ 0 & \text{if } X \neq \mathtt{A}\,. \end{cases}$$

Of course, the same notation is used with A replaced by `C`, `G` or `T`.

We observe that the evidence is probably against such a simplistic model. In genetics, `CpG` islands or `CG` islands are genomic regions that contain a high frequency of `CpG` sites. `CpG` or `CG` sites are regions of DNA where a cytosine nucleotide is followed by a phosphate link and then a guanine nucleotide. Sometimes a gene may have a `CpG` island and this implies that this iid model is not good here.

Even if this initial model is inadequate, or unrealistic, it can still serve as a **reference model** against which to make comparisons.

Appendix F
Earlier Applications of Hidden Markov Chain Models

Introduction

In this appendix some earlier application methods are briefly described.

Markov chain models can be used to provide probability models for sequences of symbols. This will aid in genome annotation. The types of questions that can be asked include the following: Does a particular sequence belong to a particular family and what can one say about its internal structure? How can one discriminate between two sequences?

Some general reviews are given in (Durbin et al., 1998, Chapters 2 and 3), (Robin et al., 2005, Chapters 1 and 2), but a more detailed review of observed Markov chains is provided by (Koski, 2001, Chapter 9). We have added some extra details to Koski's treatment.

A straightforward application of Markov chains to genome sequencing. This approach does not seem to work for the following reasons:

- The four bases A, T, G, C are not uniformly distributed in a sequence and the compositions vary within and between sequences.
- Various k-tuples of bases are not uniformly distributed. However, exons and introns are often separated on the basis of dinucleotide frequencies.
- It seems that higher-order chains need to be used as probabilities of a base in a particular location and then can depend not only on the immediately adjacent bases. In addition, the base composition can vary from one segment to another. The segmentation techniques for decomposing DNA sequences into homogeneous segments includes hidden Markov models.

Frame-dependent Markov chains. These use the GeneMark software; information can be found at

`http://genemark.biology.gatech.edu/GeneMark/gm_info.html`

Mixture transition distribution chain of order k. These are called MTD(k) models. For a Markov chain of order k with a state-space of size N, there are $(N-1)\,N^k$ entries in the transition matrix A to be estimated, (the column sums of A are 1), plus the initial probabilities. With $N = 4$ and $k = 8$, we have $3 \cdot 4^8 = 196,608$ which is quite large. This has a further implication that we may not have enough data to calibrate all these entries in A. We comment on estimation using sparse data below.

Berchtold and Raftery (2002), Raftery (1985) and Raftery and Tavare (1994) proposed a model with $N(N-1)+(k-1)$, which equals 19 in the example for the number of entries in A.

In this model there are initial probabilities $\pi_1, \pi_2, \ldots, \pi_k$ which are non-negative (the above authors relax this requirement but use it in examples.) with sum 1 and $Q = (q_{ji})$ (that is our $A = (a_{ji})$) a matrix with non-negative entries whose column sums are 1, (see Bartlett, 1978) so that the order k Markov chain $\{X_n\}$ has statistics described by

$$P(X_{n+1} = e_k \mid X_n = e_{i_1}, X_{n-1} = e_{i_2}, \ldots, X_{n-k+1} = e_{i_k}) = \sum_{r=1}^{k} \lambda_r \, q_{k,i_r}$$

The semi-martingale representation becomes

$$X_{n+1} = \sum_{r=1}^{k} \lambda_r \, Q \, X_{n+1-r} + M_{n+1}$$

where $\mathbf{E}[M_{n+1} \mid \mathcal{F}_n] = 0 \in \mathbb{R}^N$ for all $n+1 \geq k$. Of course, these two specifications are equivalent.

Raftery (1985) indicates that the model parameters are estimated using maximum likelihood estimates. A similar method to that used in Chapter 2 can be used to construct a likelihood function, by assuming some distribution for $(X_0, X_1, \ldots, X_{k-1})$ and then setting

$$\bar{\lambda}_l = N \left\langle X_l, \sum\nolimits_{r=1}^{k} \lambda_r \, Q \, X_{l-r} \right\rangle \quad \text{for } l \geq k.$$

We leave the details as an exercise. Raftery and Tavare (1994) give an alternative way of estimating the model by minimum χ^2 estimation, and provide an example for DNA analysis. Berchtold and Raftery (2002) provide further estimation techniques and give further DNA examples.

Lumping. Suppose $\{X_n\}$ is a Markov chain and $g : \mathbb{R}^N \to \mathbb{R}^M$ where $M < N$. Then $Y_n = g(X_n)$ may not be a Markov chain. Here $\{Y_n\}$ is called a lumping of the chain $\{X_n\}$.

Consider the alphabets

$$\{A, G, T, C\} \to \{\{A, G\}, \{C, T\}\}$$

Here we have $N = 4$ and $M = 2$. We associate the first alphabet with $\{e_1, e_2, e_3, e_4\} \subset \mathbb{R}^4$ and the second with $\{e_1, e_2\} \subset \mathbb{R}^2$. The function g would be given by

$$g(e_1) = g(e_2) = \begin{bmatrix} 1 \\ 0 \end{bmatrix}$$

and

$$g(e_3) = g(e_4) = \begin{bmatrix} 0 \\ 1 \end{bmatrix}$$

So, for example, if

$$X_n = \begin{bmatrix} 1 \\ 0 \\ 0 \\ 0 \end{bmatrix} \text{ or } \begin{bmatrix} 0 \\ 1 \\ 0 \\ 0 \end{bmatrix} \text{ then } g(X_n) = \begin{bmatrix} 1 \\ 0 \end{bmatrix}.$$

In this lumping `A` and `G` are lumped together as one symbol and so are `C` and `T`.

Note that

$$Y_n = g(X_n) = \sum_{i=1}^{N} g(X_n)\langle X_n, e_i\rangle = \sum_{i=1}^{N} g(e_i)\langle X_n, e_i\rangle$$

So, now the nonlinear function g of X_n is actually linear in X_n.

Lumping is also discussed in (Koski, 2001, §13.4) where a sufficient condition is given for $\{Y_n\}$ to be a Markov chain. This quotes results of Kelly (1982) who also discusses necessary conditions.

Variable length Markov chains. These are Markov chains where order of chain varies along the chain following a specified model.

Interpolated Markov models These models use a software called GLIMMER and are Markov type models which are superpositions of a variety of Markov models of various orders.

First occurrence of a word in Markov chain. Koski (2001) reports on the studies of Robin and Daubin (1999). These results are also discussed in more detail in Robin et al. (2005).

Conditional expected frequency of a word in a Markov chain. There are a variety of results for Markov chains due to Cowan (1991) Prum et al. (1995) on the frequency of DNA patterns using a so-called Whittle's formula for multinomial coefficients. Robin et al. (2005) discusses these ideas, as does Reinart et al. (2000).

References

Aggoun, L., Elliott, R. (2004). *Measure Theory and Filtering: Introduction with Applications.* Cambridge University Press.

Barbu, V.S., Limnios, N. (2008). *Semi-Markov Chains and Hidden Semi-Markov Models Towards Applications: their Use in Reliability Theory and DNA Analysis.* Springer.

Bartlett, M.S. (1978). *An Introduction to Stochastic Processes,* third edition. Cambridge University Press.

Berchtold, A., Raftery, A.E. (2002). The mixture transition distribution model for high-order Markov chains and non-Gaussian time series. *Statistical Science,* **17** (3), 328–356.

Borodovsky, M., Ekisheva, S. (2006). *Problems and Solutions in Biological Sequence Analysis.* Cambridge University Press.

Bressler, Y. (1986). Two-filter formula for discrete-time non-linear Bayesian smoothing. *Int J. Control,* **43**, 629–641.

Bulla, J. (2006). *Application of Hidden Markov Models and Hidden Semi-Markov Models to Financial Time Series.* PhD Thesis, Universität Göttingen.

Bulla, J., Bulla, I. (2006). Stylized facts of financial time series and hidden semi-Markov models. *Computat. Stat. and Data Anal.,* **51**, 2192–2209.

Bulla, J., Bulla, I., Nenadić, O. (2010). `hsmm` – An **R** package for analyzing hidden semi-Markov models. *Computational Statistics and Data Analysis,* **54**, 611–619.

Burge, C. (1997). *Identification of Genes in Human Genomic DNA.* PhD Thesis, Stanford University.

Burge, C., Karlin, S. (1997). Prediction of complete gene structures in human genomic DNA. *J. Mol. Biol.,* **268**, 78–94.

Chen, S.F., Goodman, J. (1996). An empirical study of smoothing techniques for language modeling. In *Proc. 34th Annual General Meeting on Association for Computational Linguistics,* 310–318.

Cristianini, N., Hahn, M.W. (2007). *Introduction to Computational Genomics: a Case Studies Approach.* Cambridge University Press.

Çinlar, E. (1975). *Introduction to Stochastic Processes.* Prentice–Hall.

Claviere, J.-M., Notredame, C. (2007). *Bioinformatics for Dummies.* Wiley.

Cohen, S.N., Elliott, R.J. (2015). *Stochastic Calculus and Applications.* Birkauser.

Cowan, R. (1991). Expected frequency of DNA patterns using Whittle's formula. *J. Applied Probability* **28**, 886–892.

Dempster, A.P., Laird, N.M., Rubin, D.B. (1977). Maximum likelihood From incomplete data via the EM algorithm. *J. Roy. Stat. Soc., Series B*, **39**(1), 1–38.

Deonier, R.C., Tavaré, S., Waterman, M.S. (2005). *Computational Genome Analysis: an Introduction.* Springer.

Devijver, P.A. (1985). Baum's forward–backward algorithm revisited. *Pattern Recognition Letters*, **3**, 369–373.

Durbin, R., Eddy, S., Krogh, A., Mitchison, G. (1998). *Biological Sequence Analysis: Probabilistic Models of Proteins and Nucleic Acids.* Cambridge University Press.

Durrett, R. (2008). *Probability Models for DNA Sequence Evolution.* Springer.

Elliott, R.J., Aggoun, L., Moore, J.B. (1995). *Hidden Markov Models: Estimation and Control.* Springer.

Ewens, W.J., Grant G.R. (2010). *Statistical Methods in Bioinformatics: an Introduction.* Springer.

Ferguson, J.D. (1980). Variable duration models for speech. In *Symposium on the Application of Hidden Markov Models to Text and Speech*, Institute for Defense Analyses, Princeton, NJ. pp. 143–179.

Fink, G.A. (2008). *Markov Models for Pattern Recognition: from Theory to Applications*, second edition. Springer.

Guédon, Y. (1992). Review of several stochastic speech unit models. *Computer Speech and Language*, **6**, 377–402.

Guédon, Y. (1999). Computational methods for discrete hidden semi-Markov chains. *Applied Stochastic Models in Business and Industry*, **15**, 195–224.

Guédon, Y. (2003). Estimating hidden semi-Markov chains from discrete sequences. *Computer Speech and Language*, **12**, 604–639.

Guédon, Y. (2004). Exploring the state sequence space for hidden Markov and semi-Markov chains. *Computer Speech and Language*, **51**, 2379–2409.

Guédon, Y., Cocozza-Thivent, C. (1990). Explicit state occupancy modelling by hidden semi-Markov models: application of Derin's scheme. *Computer Speech and Language*, **4**, 167–192.

Gusfield, D. (1997). *Algorithms on Strings, Trees, and Sequences: Computer Science and Computational Biology.* Cambridge University Press.

Harlamov, B. (2008). *Continuous Semi-Markov Processes.* Wiley.

Howard, R.A. (1971). *Dynamic Probabilistic Systems. Volume II: Semi-Markov and Decision Processes.* Wiley.

Ignatova, Z., Martínez-Pérez, I., Zimmermann, K.-H. (2008). *DNA Computing Models.* Springer.

Isaev, A. (2006). *Introduction to Mathematical Methods in Bioinformatics.* Springer.

Janssen, J., Manca, R. (2010). *Semi-Markov Risk Models for Finance, Insurance and Reliability.* Springer.

Jelinek, F. (1997). *Statistical Methods for Speech Recognition.* MIT Press.

Jelinek, F., Mercer, R.L. (1980). Interpolated estimation of Markov source parameters from sparse data. In *Pattern Recognition in Practice*, Gelsema, E.S., Kanal, L.N. (eds.), pp. 381–397. North-Holland.

Jones, N.C., Pevzner, P.A. (2004). *An Introduction to Bioinformatics Algorithms.* MIT Press.

Kelly, F.P. (1982). Markovian functions of a Markov chain. *Sankhya*, **44** 372–379.

Koski, T. (2001). *Hidden Markov Models for Bioinformatics.* Kluwer Academic.

Krishnamurthy, V., Moore, J.B., Chung, S.-H. (1991). On hidden fractal model signal processing. *Signal Processing*, **24**, 177–192.

Levinson, S.E. (1986a). Continuously variable duration hidden Markov models for speech analysis. In *Proceedings ICASSP, Tokyo*, 1241–1244.

Levinson, S.E. (1986b). Continuously variable duration hidden Markov models for automatic speech recognition. *Computer Speech and Language*, **1**, 29–45.

McLachlan, G.J., Peel, D. (2000). *Finite Mixture Models.*. Wiley.

McLachlan, G.J., Krishnan, T. (2008). *The EM Algorithm and its Extensions, second edition.* Wiley.

Oksendal, B. (2010). *Stochastic Differential Equations, 6th edition.* Springer.

Pardoux, E. (2008). *Markov Processes and Applications: Algorithms, Networks, Genomes and Finance.* Wiley.

Pevsner, J. (2003). *Bioinformatics and Functional Genomics.* Wiley.

Prum B,, Rodolphe F., Turckheim E. (1995). Finding words with unexpected frequencies in DNA sequences. *J. Royal Stat. Soc. Series B*, **57**, 205–220.

Rabiner, L.R. (1989). A tutorial on hidden Markov models and selected applications in speech recognition. *Proceedings of the IEEE*, **77** (2), 257–287.

Rabiner, L., Juang, B.-H. (1993). *Fundamentals of Speech Recognition.* Prentice–Hall.

Raftery, A.E. (1985). A model for high-order Markov chains. *J. Royal Stat. Soc. Series B*, **47**(3), 528–539.

Raftery, A.E., Tavare, S. (1994). Estimation and modelling repeated patterns in high-order Markov chains with the mixture transition distribution model. *Applied Statistics*, **43**(1), 179–199.

Ramesh, R., Wilpon, J.G. (1992). Modeling state durations in hidden Markov models for automatic speech recognition. In: *IEEE International Conference on Acoustics, Speech and Signal Processing, ICASSP-92, volume 1*, pp. 381–384.

Reinart, G., Schbath, S., Waterman, M. (2000). Probabilistic and statistical properties of words: an overview. *J. of Computational Biology*, **7**, 1–46.

Robin, S., Daubin, J.J. (1999). Exact distribution of word occurrences in a random sequence of letters. *J. Applied Probability*, **36**, 179–193.

Robin, S., Rodolphe, F., Schbath, S. (2005). *DNA, Words and Models: Statistics of Exceptional Words.* Cambridge University Press.

Shmulevich, I., Dougher, E.R. (2007). *Genomic Signal Processing.* Princeton University Press.

Steeb, W.-H., Hardy, Y. (2011). *Matrix Calculus and Kronecker Product: a Practical Approach to Linear and Multilinear Algebra*, Second edition. World Scientific.

van der Hoek, J, Elliott, R.J. (2013). A modified hidden Markov model. *Automatica*, **49**, 3509–3519.

Wall, J.E., Willsky, A.S., Sandell, N.R. (1981). On the fixed interval smoothing problem. *Stochastics*, **5**, 1–42.

Waterman, M.S. (2000). *Introduction to Computational Biology*. Chapman and Hall/CRC.

Yu, S.-Z. (2010). Hidden semi-Markov models. *Artificial Intelligence*, **174**, 215–243.

Yu, S.-Z., Kobayashi, H. (2003a). An explicit forward–backward algorithm for an explicit-duration hidden Markov model. *IEEE Signal Processing Letters*, **10**, 11–14.

Yu, S.-Z., Kobayashi, H. (2003b). A hidden semi-Markov model with missing data and multiple observation sequences for mobility tracking. *Signal Processing*, **83**, 235–250.

Yu, S.-Z., Kobayashi, H. (2006). Practical implemenation of an efficient forward–backward algorithm for an explicit-duration hidden Markov model. *IEEE Transactions on Signal Processing*, **54**, 1947–1951.

Zucchini, W., MacDonald, I.L. (2009). *Hidden Markov Models for Time Series: an Introduction using R*. Chapman and Hall/CRC.

Index